Design of Marine Risers With Functionally Graded Materials

Woodhead Publishing Series in Civil and Structural Engineering

Design of Marine Risers With Functionally Graded Materials

SRINIVASAN CHANDRASEKARAN

ELSEVIER

WP
WOODHEAD PUBLISHING
An imprint of Elsevier

Woodhead Publishing is an imprint of Elsevier
The Officers' Mess Business Centre, Royston Road, Duxford, CB22 4QH, United Kingdom
50 Hampshire Street, 5th Floor, Cambridge, MA 02139, United States
The Boulevard, Langford Lane, Kidlington, OX5 1GB, United Kingdom

British Library Cataloguing-in-Publication Data
A catalogue record for this book is available from the British Library

Library of Congress Cataloging-in-Publication Data
A catalog record for this book is available from the Library of Congress

ISBN: 978-0-12-823537-9 (print)

ISBN: 978-0-12-823561-4 (online)

For information on all Woodhead Publishing publications
visit our website at https://www.elsevier.com/books-and-journals

Publisher: Matthew Deans
Acquisitions Editor: Gwen Jones
Editorial Project Manager: Mariana L. Kulh
Production Project Manager: Anitha Sivaraj
Cover Designer: Matthew Limbert

Typeset by MPS Limited, Chennai, India

Contents

Preface

Design of Marine Risers With Functionally Graded Materials explores the novelty of functionally grading materials for marine risers. The constituent elements are chosen to satisfy both the strength and service life requirements of marine risers in a corrosive environment. A combination of a variety of environmental loads that arise from waves, current, and those arising from the impact of vessels will result in the development of extreme stresses. It makes the analysis and design of marine risers, a complicated process. With progress in the research and scientific developments, several geometric forms of the production risers such as flexible risers, steel catenary risers, and hybrid risers are also deployed successfully for both exploration and production. The first chapter deals with the basics of marine risers, various geometric configurations, layout preferences for offshore platforms, and fatigue load issues. Besides, alternate materials used for marine risers are also briefly discussed.

Under the corrosive marine environment, a corrosion-resistant alloy of duplex stainless steel, functionally graded with carbon–manganese steel along with its inner layer, is examined. Detailed experimental studies carried out to assess both the mechanical and metallurgical properties are discussed in the second chapter. The interface strength of functionally graded materials (FGM), fabricated using cold metal transfer, gas metal arc welding, based on the wire arc additive manufacturing in the lab scale, is compared with an X52 carbon–manganese steel to highlight the superiority of the proposed FGM.

The design of risers to withstand the oscillatory motion during the formation of vortices is a challenging task. Vortex-induced vibration (VIV) is the outcome of such exciting, periodic forces, causing a response closer to the resonance period. Being undesirable, the suppression of VIV will improve the safe and comfortable working environment of offshore drilling by enhancing the design life of structural members. In the third chapter the VIV phenomenon in the context of marine risers is presented in detail. A few recent research studies carried out to suppress VIV in cylindrical members are also discussed. Experimental investigations carried out on a scaled model to examine its behavior with and without suppression systems are presented. It is seen that by attaching stripping wires, turbulence is generated within the laminar boundary layer of the model.

Such protrusions on the surface of the bluff body have a considerable effect on the vortex-induced formation process and resulting oscillations of the cylinder.

The recent development of structural forms of offshore compliant structures is unique and expensive by design, installation, commissioning, and operability. See, for example, the buoyant leg storage and regasification platform and triceratops. Ocean structures are more focused on being form-dominant rather than strength-based design. Compliant and floating structures, when coupled with risers and moorings, show a complex behavior under environmental loads. Therefore recent studies lead to the use of modern materials for their construction as their repair is undesirable due to their limited accessibility for structural intervention. In the fourth chapter the use of different materials is discussed, which will be complementary to FGM. The dynamic analyses of semisubmersibles under the postulated failure of mooring lines, discussed in the chapter, will be an investigative research domain. Further, a novel study on offshore triceratops with buoyant legs of the elliptical cross section is also presented, which elaborates on the recent advancements in the cross-sectional shape of members of offshore platforms.

The author expresses his immense gratitude to all his teachers, research scholars, graduate students, and colleagues for their support and advice at various capacities. In particular, the research outcome shared by Merin Thomas, Syed Azeem Uddin, and Dr. Nagavinothini needs a special mention. The author also specially thanks Dr. Hari Sreenivasan, Assistant Prof., School of Petroleum Technology, Pandit Deendayal Petroleum University, Gandhinagar for coauthoring the second chapter of this book. The author also expresses his gratitude to Prof. Murugaiyan Amirthalingam, Department of Metallurgical and Materials Engineering, Indian Institute of Technology Madras for his valuable inputs and technical support. The author also thanks the Chairman, Centre of Continuing Education, Indian Institute of Technology Madras, for extending administrative assistance in preparing the manuscript of this book.

List of Figures

List of Tables

CHAPTER 1

Marine risers: introduction

1.1 Introduction

Marine risers are cylindrical conduits that are used to transfer crude oil from the subsea wellhead to the offshore facility. Marine risers are used in oil and gas exploration and production since the 19th century. Earlier use of risers can be seen in the drilling activities carried out in the offshore barges of California (Chandrasekaran and Jain, 2016; Chandrasekaran, 2015b, 2017). In the 1960s risers are also used in barges, which are dynamically positioned to enable oil exploration activities. Marine risers shall possess two distinct geometric configurations: rigid and flexible. Hybrid risers are a combination of both the flexible and rigid type of configurations. As risers are connected to the drilling and production offshore platforms (or vessels, in case of drillships), they are designed to remain flexible so that flexible movement of the host platforms is not intervened. While the choice of an appropriate geometric configuration is an essential factor, the length of the riser is also equally important, as this shall impose a direct impact on the cost of material and installation of risers. Besides, stresses in the riser cross section and connections should be within the desirable limits. A combination of a variety of environmental loads that arise from waves, current, and those arising from the impact of vessels shall result in the development of extreme stresses. Further, vortex-induced vibrations (VIVs) and fatigue cause more severe concerns on the sustainable service life of risers.

1.2 Type of risers

Primarily, risers are used for the following purposes, namely, drilling, production, injection, completion, workover, and export or import of oil and gas. Depending on the purpose and application, risers differ in their geometric configurations and material. Drilling risers, deployed for drilling oil wells in deep offshore, are classified as low pressure and high pressure based on the operational conditions. Production risers are developed for

Design of Marine Risers with Functionally Graded Materials
DOI: https://doi.org/10.1016/B978-0-12-823537-9.00006-2

transporting oil or gas, more or less, continuously from the well to the top side of an offshore platform. Production risers are of a larger diameter and usually circular in cross section and subjected to lateral loads of varying magnitude along its length. Due to their longer length and slender cross-sectional characteristics, these risers are suspended down from the platform and upheld under high axial tension; these are termed top-tensioned risers (TTRs). With progress in the research and scientific developments, several geometric forms of the production risers such as flexible risers, steel catenary risers (SCRs), and hybrid risers are also deployed successfully for both exploration and production.

1.3 Drilling riser

Drilling risers are made of riser joints, each segment ranging about 15–23 m long (Sparks, 2007). Low-pressure drilling risers consist of a central tube and several flow lines along its periphery. Fig. 1.1 shows a drilling riser with peripheral lines. Generally, four flow lines are used, kill, choke, booster, and hydraulic lines. In the event of a gas kick, kill and choke lines are used for well-control operations, shown in green and brown color in the figure, respectively. Booster lines are used to improve the flow by injecting fluid at the lower end of the riser, which enables a quicker discharge of the drill cuttings while the hydraulic lines are used in the blowout preventer (BOP) at the seabed.

It also consists of syntactic foam that enables the reduction in weight of the riser in water. The syntactic foam is generally seen in the upper sections of the marine riser. The section closer to the surface is usually left without any foam to reduce the hydrodynamic loads under the encountered waves. Air-can buoyancy modules were attached in the earlier design of drilling risers. Their buoyancy was varied depending on the drilling requirements. Different configurations of the connectors are used for the riser joints such as breechblock and flanged connector. Flex joint, placed between the upper and lower sections of the riser, permits a finite rotation of the riser. It is useful in case of an emergency. The BOP shears the drill string to keep it closed. High-pressure drilling riser is used when the BOP is housed on the surface of the subsea facility; its configuration does not possess kill and choke lines. High-pressure risers are used to drill wells in many of the tension leg platforms (TLPs), such as Hutton, Heidrun, and Mars.

Figure 1.1 Drilling riser with peripheral lines.

1.4 Completion and workover risers

Both the tubing and its hangers of a completion riser shall run into the wellbore, through the BOP. Loads on these risers arise due to the curvature of the riser and the joints at both ends. To recenter the well through the subsea tree, a workover riser is deployed. They are quite similar to that of the drilling risers. The completion and workover risers do not consist of choke and kill lines. A wellhead is housed on the platform, which is designed to sustain a high pressure. However, there is provision for an emergency disconnect when they are used in a mobile offshore drilling unit; it is termed an emergency disconnect package. However, for sealing the well at the seabed, a riser safety package is deployed. Umbilical powers these packages and are also attached to the risers.

1.5 Top-tensioned risers

As the name suggests, TTR depends on the top tension imposed on the riser, which is significantly higher than its apparent weight. Fig. 1.2 shows a schematic view of a TTR. TTR is initially installed in the Hutton TLP and, subsequently, found their extensive applications in both TLPs and single point anchor reservoir (SPAR) platforms. Wellhead is located on the offshore platform, which extends direct access to the oil well. The stress joint is used for connecting the riser to the seabed, which confirms a flexible connection. TTR is designed to withstand the tubing pressure, in case of any failure in the tubing. Keel joint is used in the case of SPAR to

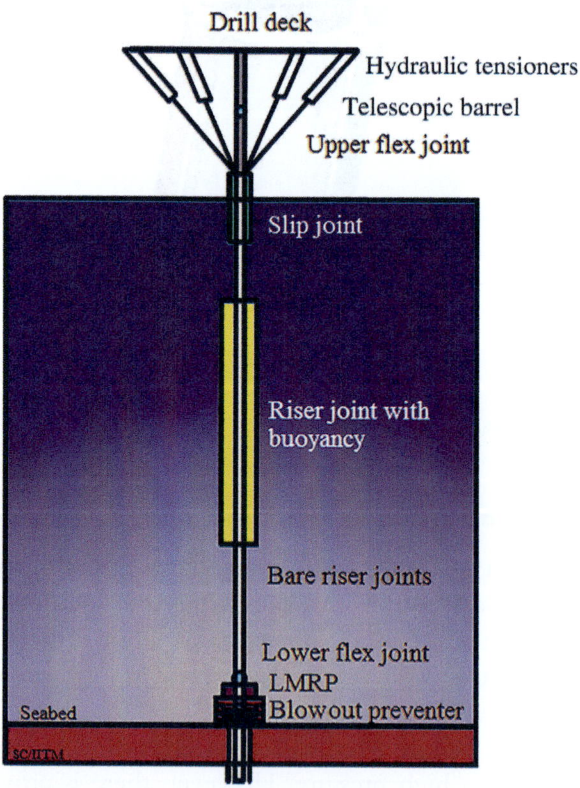

Figure 1.2 Schematic views of a top-tensioned riser.

release the additional bending stress caused under the failure of tubing. It is generally seen that the TLPs use tensioners while SPAR uses air cans.

1.6 Steel catenary risers

In the presence of wave and current loads, rotation at both the upper and lower ends of a riser is limited in the case of TTRs, imposing a severe limitation. But, in the SCR, heave motion compensates for the loss of tension. Fig. 1.3 shows an SCR attached to a semisubmersible. In the presence of strong currents a reduction in the top tension can result in a very high bending moment along the length of the riser. Further, Euler buckling can also occur in case if the active tension is negative. An SCR can overcome all these disadvantages. An SCR can self-compensate for the heave movement, by its geometric profile. Auger TLP is one of the first offshore platforms where SCRs are deployed as export risers in the year 1994. The use of SCR has multiplied in the coming years to several offshore facilities as production risers. One of the most significant advantages of the SCRs is that they are capable of withstanding fluctuations in tension, fatigue, and movement at the touchdown point. Ball joints, provided in the connections, permits rotation at the joints and offer a significant advantage under the combined action of waves, vessel motions, and currents. However, the effects of VIVs can cause severe consequences to an SCR. Helical strakes, provided at specific sections along the length of the riser, are capable of suppressing the VIV, to a significant level.

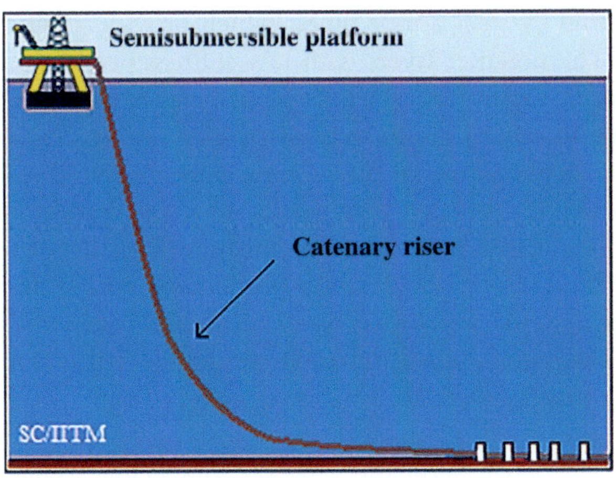

Figure 1.3 Catenary riser attached to a semisubmersible.

An API grade steel of X-52 or higher is commonly used in SCR. VIV results in an amplification of the drag, thereby causing an increase in the bending stress. Besides, crashing loads that occur due to the interaction between risers are also critical in the design of SCR. Further, the laying process and its effect on the welding of risers result in uncertainty in the plastic strain that is one of the critical design considerations of SCR. In the deepwater environment, clay lies loosely on the seabed. With time, due to the riser—soil interaction, riser pipe sinks into the loose clay on the seabed and gets buried in it. Under such situations, soil properties become essential to ascertain the soil uplift and sideways resistance. They are vital to model the interaction between the soil and the marine riser. According to the design codes, in storm conditions, the maximum equivalent stress in an SCR is limited to 80% of its yield stress. Under the anticipated conditions of a failure of the mooring line or tether, 100% yield stress is tolerable. A similar approach is used in risers that are vertically tensioned falling under the ASME boiler code and API RP 2RD.

1.7 Flexible risers

Flexible risers were introduced to the offshore industry in the 1970s, with their initial deployment in shallow waters; it is later extended to even high-pressure applications. Flexible riser comprises several concentric pipes, which results in lower bending stiffness in comparison to its axial stiffness. Low bending stiffness is that when pipes slip against one another under the action of the internal and external loads. However, this weakness is overcome by steel armor pipes, which provide additional strength and stiffness. The polymer layers with low stiffness provide fluid integrity. One of the most exceptional advantages of flexible risers is that they can are deployable in several configurations of riser layout. Further, they can be stored in reels, which make them convenient for transport and installation. Therefore they are cost-effective in terms of installation, transport and are highly suitable for compliant offshore structures such as TLPs, SPAR, semisubmersibles, and floating production storage and offloading. Various configurations are chosen based on the requirement in their production and site-specific environmental conditions. Fig. 1.4 shows different layers of a flexible riser.

A flexible riser is a tube made of a flat strip of stainless steel, the grade of which largely depends on the type of fluid being conveyed through the riser. The carcass is the innermost layer of a flexible riser, which offers

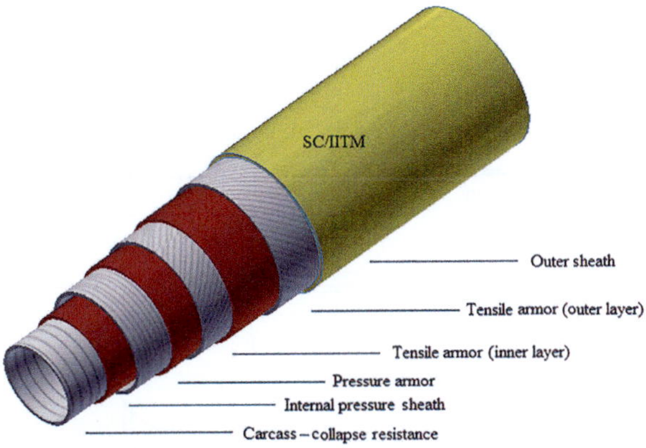

Figure 1.4 Different layers of a flexible riser.

resistance against the collapse of the riser under high hydrostatic pressure and gases that build-up in their annulus. In the event of depressurization, which occurs after a well is shut, very high internal stresses are developed; the steel carcass layer is capable of withstanding this collapse pressure. An inner sheath forms the second layer, which is made of high-density polyethylene (HDPE); cross-linked polyethylene is also used as alternatives. This polymer sheath provides bore-fluid integrity. The third layer is the pressure armor, which is designed to withstand the hoop stress generated in the inner bore of the riser. They are wound over the internal sheath and made of interlocking wires. Carbon steel of high strength is commonly used as pressure armor. Tensile armors are made of carbon steel of high strength and are capable of resisting tensile loads in the risers. They are made of rectangular wires that are flat and cross-wound at an angle to the axis of the riser. For deepwater applications the number of layers of tensile armors is usually more; this is required to withstand the high tensile load. The external sheath is made of HDPE or cross-linked polyethylene, similar to that of the internal sheath. It forms a protective layer and prevents seepage of seawater into the riser. During the process of installation the external sheath acts as a protection for the internal layers of the riser.

1.8 Layout configurations

Six main configurations are available for a flexible riser, as shown in Fig. 1.5. They are, namely, free-hanging catenary, lazy-wave, steep-wave, lazy-S, steep-S, and pliant-wave configurations. Free-hanging catenary

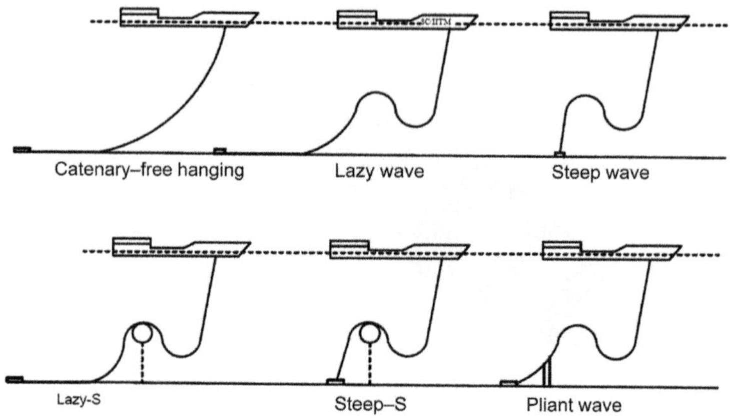

Figure 1.5 Riser layout configurations.

layout configuration is one of the simple and cheapest configurations of a flexible riser owing to its simple installation and minimum requirements of infrastructure at the subsea. It undergoes high loading under the movement of the connecting vessel or offshore platform. Under high motions of the vessel, free-hanging catenary can be subjected to compression buckling at the touchdown point. It can result in the formation of *bird-caging* in the tensile armor wire (Bai and Bai, 2005). When used in deeper waters, a higher magnitude of top tension is provided to take care of the extra length of the riser that needs to be supported.

Both in the lazy-wave and steep-wave layouts, vessel motions are decoupled from its effects on the touchdown point by adding buoyancy and weight along with the remaining (longer) length of the riser. Lazy-wave configuration requires lesser infrastructure in the subsea and is preferred more over the steep-wave layout. A lazy-S layout needs an alternate configuration for transporting fluid with a change in density; steep-wave risers can still maintain the same configuration. Both base and bend stiffeners are used at intermittent intervals, in regular spacing in the steep-wave risers to support the layout of risers. Syntactic foam is used in the buoyancy modules due to its low water-absorption properties. High stress may be developed in the armor wires in the riser in case of any slippage of these buoyancy modules. Clamps, provided with the buoyancy modules, help to avoid slippage failure but may also cause additional wear and tear to the external sheath; it may subsequently result in water ingression in the annulus. However, these risers are designed to handle about 10% loss in the buoyancy, which may occur in due course their service life.

Lazy-S type and steep-S type of flexible riser configurations consist of a subsea buoy, which can either be fixed to the offshore platform or position restrained by moorings. Subsea buoys, provided with these risers, eliminate the problems faced by the risers at the touchdown point. Further, they are also advantageous in handling the additional tension imposed on the risers under the severe wave environment. Only a marginal amount of this additional tension is transferred to the touchdown point. These types of layout configurations are used if catenary geometry is not suitable due to the complexities that arise during their installation. Tether, its base, and a midwater arch are required in the lazy-S type of flexible riser, whereas bend stiffener and buoy are required in the steep-S type of configuration. In the case of vessels with more significant motion, a steep-S is highly desirable due to the compression load developed at the touchdown zone.

Pliant-wave configuration is very similar to that of the steep-wave type. In this type, there are anchors at the seabed, which handle uplift motion that occurs at the touchdown point. This arrangement enables the tension from the riser to be transferred to the subsea anchor and thus releases the tension from the touchdown point. One of the significant advantages of this type of configuration is that the riser can handle a wide range of fluid with varying densities. It also develops excellent compatibility with the significant motion of the vessel or the offshore platform to which it is connected. Therefore there is no need to alter the existing configuration, as required in the earlier types of risers; it also avoids the development of higher internal stresses in the riser. It is only used when the three configurations of catenary, lazy, or steep waves are not feasible.

1.9 Riser components

A marine riser should be able to withstand both axial tension and bending moments while remaining flexible to withstand the fatigue loads. Following are a few critical components of a marine riser system:

1. Buoyancy modules: They are attached to the riser closer to the mean sea level to reduce tension at the surface. Generally, air cans or syntactic foams are attached to the riser. However, the choice of material for buoyancy modules largely depends on their water-absorption capacity and resistance to impact loads.
2. Riser joints: They are manufactured of a seamless pipe wherein the connectors are welded to its ends. Peripheral lines such as kill, choke,

and hydraulic are attached to the riser with the help of flanges extending from the connector.

3. Bell mouths and bending stiffeners: They are critical components of a marine riser and attached to the top section of the riser, well before the hang-off accessory. It is important to note that this section of a riser is highly prone to bending. Bell mouths and bending stiffeners help to reduce the excessive bending stresses by reducing the bend radius and also by increasing the stiffness of the riser. In particular, bending stiffeners are used at sections where the vessel motion is high. The load imposed on the flexible risers is used as an input to design these components. Polyurethane is used to manufacture the bending stiffeners as it is capable of imparting a gradual increase in the stiffness to the particular section of the riser. They are anchored with a steel collar to enable effective transfer of load. Both the accessories are also used in steep-S and steep-wave layout configurations at subsea locations. They provide excellent end support, besides, to prevent it from overbending. Bell mouths offer a similar function as that of a bending stiffener and are fabricated using steel. They are manufactured with high tolerances to prevent any formation of minor cracks leading to stress concentrations.
4. Bend restrictors: They are connections used at the top and bottom of a riser to restrict the bending radius of the riser under lateral loads. Additional resistance is provided from the overbending of the riser at the critical locations by mechanical locking. They are fabricated using plastic materials that are hard enough to resist excessive bending. They are also used in the base of risers and wellheads (Bai and Bai, 2005).

1.10 Vortex-induced vibrations

A riser is connected to a BOP valve of the wellhead to prevent any leakage during the transportation of oil and gas. As the risers vibrate under lateral loads that arise from waves and current, vortices along the surface of the riser are generated. It results in the formation of an unstable wake zone around the riser, which is highly unstable. Further, shedding of vortex occurs at varying amplitudes and frequencies, affecting the vortex patterns. While the Reynolds number of the incident flow decides the pattern of the vortices formed (Xu et al., 2009), they impose an additional transverse force of varying amplitude on the riser; it is also periodic. These forces act normal to the flow direction, resulting in VIVs (Wu et al.,

2015). The phenomenon of VIV is highly complex as it shifts continuously between the chaotic and stationary states (Modarres-Sadeghi et al., 2011). There are mainly three patterns of vortices formed: 2S, 2P, and P + S, as shown in Fig. 1.6.

A lock-in condition sets in when the frequency of vortices matches the natural frequency of vibration of a riser. It results in vibrations of a very high magnitude. It can cause severe damage to the riser. Besides, risers are also subjected to fatigue damage under the wave and current loads that are varying continuously both in their magnitude and direction of flow (Xue et al., 2008, 2014, 2018; Hong and Shah, 2018; Jeans et al., 2002; Elosta et al., 2014; Zhu and Collette, 2017; Zhang and Tang, 2015). One of the significant concerns about the structural integrity of a riser is fatigue. It arises under two conditions: (1) in the direction of flow, termed inline forces, and (2) perpendicular to the flow, termed transverse forces (Xue et al., 2014; Govardhan and Williamson, 2002). While the effects caused by inline forces are minimum in comparison to that of the latter, both are important for fatigue analysis of risers (Lie and Kaasen, 2006; Liu et al., 2016; Leira et al., 2005; Trim et al., 2005; Tim Finnigan et al., 2005). In the lower modes of vibration, fatigue damage contribution primarily arises from the inline forces, while the transverse forces

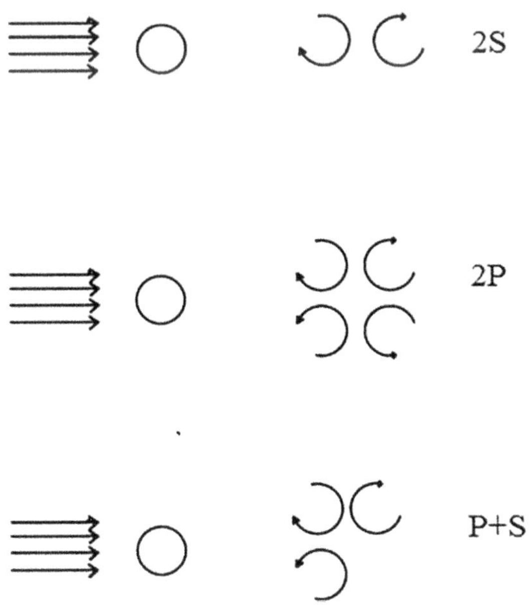

Figure 1.6 Vortex shedding patterns formed during flow on a cylinder.

contribute to the damage in higher modes of vibration (Baarholm et al., 2006; Anil, 2014; Abbassian and Moros, 1996; Chakrabarti, 2002). It has also been observed that the maximum damage occurs at the riser ends (Gao et al., 2011, 2015; Gouveia et al., 2015; Zhang and Tang, 2015; Chakrabarti and Frampton, 1982).

The hydrodynamic loads acting on the riser and the damage induced by the motion of the vessel are of significant concern. A more significant relative motion between the riser and the connected vessel or an offshore platform can challenge the safety of the riser and the production system in total. The heave motion of the vessel can cause instability due to the variations of tension in the riser. It can also result in the dynamic buckling of the riser locally. Large lateral movements of the vessel (or the complaint platform) can result in the variation of contact angles of the riser with the BOP. If the end angle increases beyond a threshold value (2 degrees), the connection between the riser and the BOP could be separated (Nguyen et al., 2010; Niedzwecki and Liagre, 2003; Pankanin et al., 2000). It is, therefore, essential to suppress these vibrations and protect the riser from an unprecedented failure. Therefore both the issues of the instability as well as fatigue have to be addressed. It is observed that in deep sea the damage caused by the vessel movement is lesser owing to its longer length when compared to the force acting on the riser due to the current (Trim et al., 2005; Yan et al., 2018; Yang et al., 2016; Yuan et al., 2011). However, in shallow depths, one of the main concerns is the instability caused due to the vessel motion. As one of the passive methods of control, the vessel is anchored firmly to the seabed using anchor lines.

However, yaw and pitch motions are regulated by providing thrusters that can regulate the motion of the vessel in the desired direction (How et al., 2009; Hover et al., 2001; Huang and Dareing, 1968; Huang et al., 2011; Hussain et al., 2009). One of the methods to prevent fatigue damage is to consider a higher factor of safety for the riser, which is also a function of water depth (Leira et al., 2005). For the suppression of VIVs, strakes and spoilers are attached to the surface of the riser, which subsequently improves its fatigue life as well (Gao et al., 2015; Abbassian and Moros, 1996). Several effective methods are also used to control the VIVs. A few are adaptive control (Hong and Bentsman, 1994a,b), sliding mode control (Ngo and Hong, 2010, 2012), boundary control (Kim and Hong, 2009), optimal control (Mazzilli and Sanches, 2011; Moo-Hyun, 2013), proportional–integral–derivative control (Jensen et al., 2010), and proportional–integral control (Rustad et al., 2008; Salama et al., 1998, 2002; Rodolfo et al., 2011; Qiao et al., 2014; Phifer et al., 1994).

1.11 Dynamics of risers

A marine riser is modeled as the Euler–Bernoulli beam under environmental loads. The riser is pin-jointed to the BOP and the vessel. One of the initial studies carried out on the buckling and lateral free vibration of a vertical, submerged pipe is given by Huang and Dareing (1968). For obtaining the drilling pipe lateral dynamics, the following equation was developed:

$$EI\frac{\partial^4 y}{\partial x^4} - \frac{\partial}{\partial x}[(w - w_m)x - F] + \rho\frac{\partial^2 y}{\partial t^2} = 0 \tag{1}$$

where EI is the flexural rigidity, w is the weight of the pipe per unit length, ρ is the total mass per unit length of the pipe and mass of the fluid participating in the motion, y is the lateral displacement at a position x, w_m is the weight of the drilling fluid per unit length, and F is the vertical force acting at the bottom of the pipe. Later, this equation of motion is modified to include the hydrodynamic forces, both in the vertical and horizontal directions (Chakrabarti and Frampton, 1982):

$$\frac{\partial}{\partial z^2}\left[\mathrm{EI}(z)\frac{\partial^2 y}{\partial x^2}\right] - \frac{\partial}{\partial z}\left[T_e(z)\frac{\partial x}{\partial z}\right] + m_x(z)\,\ddot{x} = f_{xs}(x, z, t) \tag{2}$$

where $T_e(z)$ is the effective tension; EI is the flexural rigidity of the riser; z is the global vertical axis with the origin at the mud line; m_x is the mass of the riser, and its contents; and f is the force density. Further modifications include both linear and nonlinear damping. Equation of motion for the lateral deflections is given by (Niedzwecki and Liagre, 2003):

$$\mathrm{EI}\frac{\partial^4 v}{\partial x^4}(x,t) - T_e(x,t)\frac{\partial^2 v}{\partial x^2}(x,t) - w(x,t)\frac{\partial v}{\partial x}(x,t) + M(x)\frac{\partial^2 v}{\partial t^2}(x,t) + c(x,t) + \left[C_D\frac{\rho_w D}{2}\sqrt{\frac{8}{\pi}}\sigma_r\right]\frac{\partial v}{\partial t}(x,t) = f_L(x,t) \tag{3}$$

where EI is the flexural rigidity; T_e is the effective tension; w is equal to the product of the mass per unit length, and acceleration due gravity for the elements; $c(x,t)$ linear viscous damping parameter; $v(x,t)$ is the horizontal displacement of the marine riser; C_D is the drag coefficient that is a function of Reynolds number; D is the outer marine riser diameter; f_L is the applied loading; σ_r is the root mean square of the relative velocity; and ρ_w is the density of water.

1.12 Corrosion in risers

The marine environment is corrosive that can cause severe degradation of the material and strength of the structural members. Material for construction should be carefully chosen so that the service life of the structural members is guaranteed. Repair of ocean structures in general and risers, in particular, is a multicomplex phenomenon not because of its limited accessibility for repair, but, due to undesirable intervention of the structure for repair and rehabilitation. As marine risers are subjected to a variety of loads, material for fabrication of risers should possess certain specific properties to sustain these loads and their combinations.

Material strength degrades under the effects caused by environmental loads. Few of them are chemical, fatigue, stress concentration, corrosion effects, and biofouling effects. It is desired that materials must have properties that ensure survivability in case of any accidents (collision). In case of excessive loads during hurricanes, they should be able to withstand high hydrostatic pressure. Given different combinations of forces acting on the risers, it is necessary to understand the essential characteristics that materials should possess to be qualified for the construction of marine risers. There exists a close relationship between the selection of material and the geometric form of the risers and layout design. Factors those become vital for selection of material include the physical and chemical properties of the materials, cost, and availability in large quantities and the availability of fabrication facilities for the chosen shape, size, and geometry.

1.12.1 Corrosion

Corrosion causes deterioration of material by chemical interaction with the environment; it generally refers to metals, but not uncommon in materials such as plastics, concrete, and wood. Corrosion results in the formation of new and less-desirable material from the original metal.

It results in a loss of function of the component or the system. One of the popular products of corrosion is rust that is formed on the steel surface. The primary corrosion cell is shown in Fig. 1.7 (Chandrasekaran and Jain, 2016). The primary corrosion cell needs three components: an anode, a cathode, and an electrolyte medium.

There is a measurable DC voltage, which can read in the metallic path between the anode and the cathode. When both the anode and cathode are electrically bonded, an anode is positively charged, and the latter is negatively charged. Like the conventional current, current discharges from the anode and is picked up at the cathode through the electrolyte. Current returns from cathode to the anode through an electrical path. This flow has a detrimental effect on the anode known as corrosion. *It is important to note that corrosion occurs at the anode and not at the cathode.* The corrosion process results in the formation of a new material that may react again or could be protective of the original metal. Anode and cathode in a corrosion process may be on two different metals connected, forming a bimetallic couple (galvanic couple) or, as in the case of rusting of steel, they may be formed on the metal surface. It is a prevalent problem in the joints of marine risers, as the coupling at the neck is manufactured with a different metal.

Steel is one of the most preferred materials for marine risers. It is unfortunate to note that steel is also the most suspected candidate for corrosion. Corrosion is primarily governed by the oxygen content present in seawater. Corrosion is governed by the rate at which a ferrous corrosion product is leached from the barrier of the film of rust. Marine risers being highly flexible members and run the water depth, their probability of

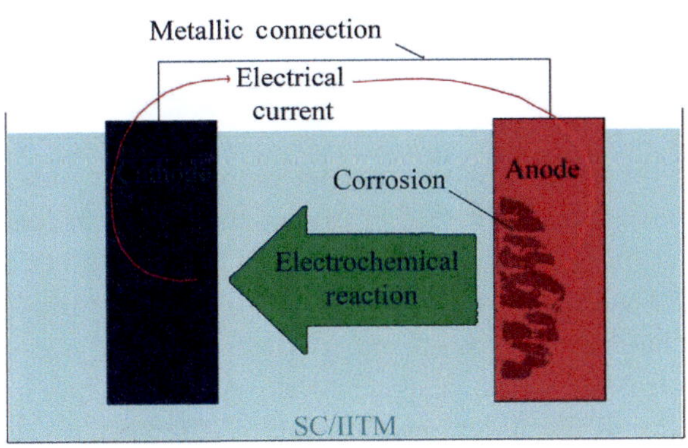

Figure 1.7 Basic corrosion cell.

corrosion severity is very high. When one of the products of corrosion becomes soluble, the formation of a protective barrier film becomes impossible. Alloys used in riser material proved to be effective in delaying the corrosion process, but unsuccessful in offering protection against corrosion. The presence of copper and nickel, even in small quantities in the low-alloy steel, enhances their corrosion resistance by altering the structure of the barrier film formation. They help to produce a tighter, denser barrier film with a lesser tendency to be removed by leaching. Severe corrosion is seen in the splash zone due to continuous wetting and drying because of tidal variations in the sea. It results in pitting corrosion in the tidal area. Due to continuous contact of seawater with a higher lateral force, thin barrier film, even if formed, shall be washed away immediately; this expedites corrosion in the splash zone. Rust films in this zone have a little opportunity to become dry as this zone is subjected to alternate wetting and drying continuously. Even though the rust films may be formed, they shall be leached off automatically. It is aggravated because of the presence of abundant oxygen content in this region. The rate of corrosion in this splash zone is several times greater than that of the continuous immersion part of the member.

A section of the steel riser in a tidal zone is in contact with highly aerated seawater and therefore becomes the cathode. As only the anodic part of the member corrodes, the adjacent submerged surface of the riser, where the oxygen content is less becomes anodic, it gets corroded severely. Current flows from anodic, which is the submerged surface of the riser to cathode section, which is the tidal zone. It demands the necessity of cathode protection to the risers in the tidal zone. In the immersed zone, corrosion in steel risers is significantly lesser due to the decrease in oxygen concentration. In this zone, corrosion is principally governed by the rate of diffusion of oxygen through layers of rust and marine organisms; the seawater temperature and tidal velocity do not influence the corrosion rate.

Further, at the touchdown point, the corrosion rate may go up in the vicinity of the mud line, but further down, they are very less. It is due to the presence of marine organisms, which can generate additional concentration cells and sulfur compounds in the vicinity of the mud line. Due to this, they become anodic compared to the remaining part and get corroded. The corrosion rate reduces well below the mud line because of the lower availability of dissolved oxygen content. Barrier films, once they are formed in this region, are relatively undisturbed. Therefore they form a protective coating automatically, and that protects the members in this region.

The effect of water depth on the corrosion rate is also an interesting viewpoint. At water depth higher than 1800 m, the temperature drops to lesser than 4°C in comparison to 24°C at the surface. It results in a substantial reduction in the corrosion rate. Metal surfaces are relatively free of marine biofouling below 700 m. Dissolved oxygen drops along with the depth but rises again below 820 m. This raise is significant in comparison to surface concentration. As risers run through the entire water depth, some sections of the risers become cathodic. Corrosion sets in at these sections at the local level along the length of the riser. It is called a *long-line effect.* Mainly due to the typical oxygen cell concentration attack, extended length of risers is subjected to different layers of varied oxygen concentration. It alters the rate of corrosion in different segments along the length of the riser, which is a long-line effect. Galvanized mooring lines are typical candidates for such problems.

1.13 Materials for risers

Under the given wide choice of available materials, the selection of material for marine risers becomes very important. Apart from the strength requirements, environmental issues related to their recycling characteristics, sustainability, renewable property, toxic, and nontoxic nature are a few essential areas of interest to offshore engineering professionals. During the selection of materials for marine risers, the following factors are generally accounted for, namely, (1) physical and chemical properties of materials, (2) cost, (3) fabrication facilities, and (4) maintenance cost. Chosen materials should avoid catastrophic failure. Besides meeting the design requirements, they should withstand hazards, including those, arise during installation and operations. The material chosen for marine risers should be easy to fabricate and transport to the offshore site, which is one of the significant constraints in material selection. It is also important to note that the expected maintenance must be looked upon as the survivability of these kinds of structural systems in the complex ocean environment is very important. Hence, the selection of material is strictly governed by the advice or recommendations given by various international codes. These codes indicate a variety of engineering properties that guide the selection of material for marine risers.

1.13.1 Steel

The most conventional material for marine risers is the steel of grade X-52 or higher. As evident, steel is one of the structural materials with a very high tensile strength and mass density. With the recent advancements in offshore drilling, which is heading toward deepwater and ultradeepwaters, riser materials are also required to sustain high pressure and temperature in addition to the lateral loads imposed by waves and current. Corrosion is one of the significant concerns in risers as the gas or oil, transported through risers, is rich in hydrogen sulfides and chloride concentration. Therefore the current focus of the offshore industry is to develop risers of lesser weight and corrosion resistant without compromising safety and durability. It has led to the use of alternate materials for marine risers. Titanium and composites are considered to be favorable alternates. Titanium alloy found its first use in the drilling riser of the Heidrun TLP. It was used for the body of the riser pipe, flanges, fasteners, and the booster lines. Titanium possesses only half the density of that of steel, which in turn reduces the weight and tension requirements of the riser. It has also resulted in compact dimensions of the flanges, while flex joints were eliminated.

1.13.2 Titanium

The titanium also offers a high yield strength that results in the reduction of the wall thickness and weight of the riser. Elastic modulus of titanium is about half of that of steel, making it an unfavorable as there is a significant reduction in stiffness. However, flex joints were preferred to reduce the effect of bending moment on the connections. The titanium also shows a high durability and corrosion resistance to seawater (Ma et al., 2015; Madec et al., 2018). The titanium also exhibits a better fatigue property in comparison to steel with a better endurance limit (Sauer et al., 1996). Besides, composite risers were also tried for the first time in the Heidrun TLP in 2001. A full-size composite drilling riser joint was used to drill three wells located at three different positions along the riser string. This demonstration led to the commercial use of riser for deepwater applications (Chandrasekaran et al., 2015a). While the composite riser, attempted in the Heidrun TLP, invoked their use in hydrocarbon industries, it also verified strong applicability to deepwater riser systems (Chandrasekaran and Mayanak, 2017). Of course, joints of composites are complicated to fabricate and commission. A computer-controlled filament welding technique was used for the fabrication of riser joints. Forty layers of carbon fiber and carbon/glass fiber were applied in the radial and longitudinal directions (Salama et al., 1998, 2002).

1.14 Moorings

Design and deployment of mooring and pipelines in deepwaters are the recent focus of analytical studies; further, improvements in materials and methodologies are also becoming equally crucial in the perspective of oil industries. The oil industry is exploring into deep and ultradeepwaters moving for which more flexible and floating structures are being deployed (Chen et al., 2011; Kurian et al., 2010). For floating structures a large pipe is required in an unusual configuration hanging below the vessel, while the vessel itself is moored using a single-point mooring. For an offshore plant the mooring and riser dynamics play essential roles (Srinivasan et al., 2006; Chakrabarti, 2005). As discussed in the earlier section of this chapter, moorings systems are generally divided into two groups: compliant and rigid. The first variety comprises catenary and taut moorings, while the latter consists of a rigid type, also termed vertical moorings. One of the essential design requirements is that a mooring line is not expected to transfer any moment. In such cases the station-keeping moment should be counterbalanced by a vertical force—couple. One of the fundamental differences between the catenary and the taut moorings is how this station-keeping moment is balanced. In the catenary moorings, compliance for wave frequency motion is ensured by a combination of geometrical and axial elasticity of the lines. The significant geometrical variations make these systems very susceptible to dynamic effects. For a taut mooring line the compliance to allow for wave frequency motion is generally provided by the axial elasticity only. As a result, the taut mooring systems do not experience substantial transverse geometric changes. The motion of the vessel, mooring, and riser interacts with each other, and their coupling effects can be classified as forces due to stiffness, damping, inertia, and external excitation. While a good design ensures that horizontal stiffness imposed by the mooring on the vessel is appropriate, some optimization is required to also account for the stiffness contribution that arises from the riser. Single-point mooring comprises straightforward configurations with a float, one mooring line, one anchor, and some additional equipment. Even for this simple system, the analytical solutions are complex. As the water depth increases, steel spread-mooring systems show several disadvantages: lower restoring efficiency, large mooring radius, and seafloor footprint. Fiber ropes are becoming popular. Polyester fiber ropes are increasingly used in deepwater mooring rigs. Large companies such as Petrobras have carried out extensive R&D on polyester ropes in MODU moorings and showed that polyester ropes could be handy for moorings in water depths higher than 500 m.

In the recent past, floating offshore platforms are culled for deepwater applications due to their better geometric form, stability trait, and construction/installation procedures. Damping, induced by mooring lines with a higher pretension, reduces with the increase in drag (Webster, 1995). Further, it is also observed that dynamic tension variations in the spread mooring and their angle of inclination could result in dynamic instability (Garza-Rios et al., 2000; Kurian et al., 2010). Several parameters such as pretension, the configuration of the mooring lines, number of mooring lines, and tension–dip angle influence the coupled motion of the platform, but the angle of inclination is found to be more effective (Chen et al., 2011; Srinivasan et al., 2006). For taut-moored conditions the geometry of a mooring layout influences the fatigue failure of mooring lines significantly; failure of one of the lines initiates failure in the adjacent ones (Huang et al., 2011; Yuan et al., 2011). It is also observed that the fatigue damage is also location dependent; it is most likely to occur near the fairlead position but shifts to the lower end under low-frequency excitations (Wu et al., 2015; Qiao et al., 2014). For different mooring configurations the emphasis is also on the fatigue analysis of intermediate sections of the mooring line (Yang et al., 2016). Studies in the recent past also showed that the fatigue damage obtained by T–N curve is more conservative than those obtained by the S–N curve (Xue et al., 2018; Yan et al., 2018).

Exercise

Part A: Objective-type questions

1. Marine risers are __________ (cylindrical conduits) that are used to transfer crude oil from the subsea wellhead to the offshore facility.
2. How extreme stresses are developed in marine risers?

   A combination of a variety of environmental loads that arise from waves, current, and those deriving from the impact of vessels shall result in the development of extreme stresses.
3. Name a few risers used for exploration and production.

   Production risers such as flexible risers, steel catenary risers, and hybrid risers are also deployed successfully for both exploration and production.

4. A riser is connected to __________ (blowout preventer) to prevent any leakage during the transportation of oil and gas.
5. How are vortices generated in risers?

   As the risers vibrate under lateral loads that arise from waves and current, vortices along the surface of the riser are generated. It results in the formation of an unstable wake zone around the riser, which is highly unpredictable.
6. What is the most fundamental aspect of the design of risers?

   As risers are connected to the drilling and production offshore platforms (or vessels, in case of drillships), they are designed to remain flexible so that flexible movement of the host platforms is not intervened.
7. List two factors that govern the design of risers.

   The choice of an appropriate geometric configuration is an essential factor. Further, the length of the riser is also equally important, as this shall impose a direct impact on the cost of material and installation of risers.
8. __________ and __________ (vortex-induced vibrations, fatigue) cause more severe concerns on the sustainable service life of risers.
9. What is the primary purpose of deploying risers?

   Primarily, risers are used for the following purposes, namely, drilling, production, injection, completion, workover, and export or import of oil and gas.
10. __________ (drilling) risers are deployed for drilling oil wells in deep offshore.
11. Name the four flow lines that are commonly used.

    Generally, four flow lines are used: kill, choke, booster, and hydraulic.
12. Where are syntactic foams used in risers?

    It also consists of syntactic foam that enables the reduction in weight of the riser in water. The syntactic foam is generally seen in the upper sections of the marine riser.
13. What is an emergency disconnect package?

    The completion and workover risers do not consist of choke and kill lines. A wellhead is housed on the platform, which is designed to sustain a high pressure. However, there is provision for an emergency disconnect when they are used in a mobile offshore drilling unit; it is termed an emergency disconnect package (EDP).

14. In the steel catenary riser, __________ (heave) motion compensates for the loss of tension.
15. ____________ (helical strakes), provided at specific sections along the length of the riser, are capable of suppressing the vortex-induced vibration.
16. According to the design codes, in storm conditions, the maximum equivalent stress in an SCR is limited to ____________ (80% of its yield stress).
17. Name a few layout configurations of a flexible riser.
    They are free-hanging catenary, lazy-wave, steep-wave, lazy-S, steep-S, and pliant-wave configurations.
18. How is the layout of risers supported along their length?
    Both base and bend stiffeners are used at intermittent intervals, in regular spacing in the steep-wave risers to support the layout of risers.
19. ____________ (syntactic foam) is used in the buoyancy modules due to its low water-absorption properties.
20. What is the most significant advantage of a steep-wave type layout of risers?
    One of the significant advantages of this type of configuration is that the riser can handle a wide range of fluid with varying densities.
21. ____________ (bend restrictors) are used at the top and bottom of a riser to restrict the bending radius of the riser under lateral loads.
22. What is the primary function of a blowout preventer?
    A riser is connected to a blowout preventer (BOP) valve of the wellhead to prevent any leakage during the transportation of oil and gas.
23. A ________ (lock-in) condition sets in when the frequency of vortices matches the natural frequency of vibration of a riser.
24. ___________ (material strength) degrades under the effects caused by environmental loads.
25. What does a primary corrosion cell need?
    The primary corrosion cell needs three components: an anode, a cathode, and an electrolyte medium.
26. Under different zones of corrosion the rate of corrosion in __________ (splash zone) is several times greater than other zones.
27. What is a long-line effect?
    As risers run through the entire water depth, some sections of the risers become cathodic. Corrosion sets in at these sections at the local level along the length of the riser. It is called a long-line effect.

28. The fatigue damage obtained by ________ (T–N curve) is more conservative than those obtained by the S–N curve.
29. Damping, induced by mooring lines with a higher pretension, reduces with the increase in ___________ (drag).
30. It is also observed that the fatigue damage is most likely to occur near the fairlead position but shifts to the lower end under ___________ (low-frequency excitations).

Part B: Descriptive-type questions

1. What are marine risers? Explain various configurations of marine risers that are commonly deployed.
2. List various types of risers that are deployed in oil and gas exploration and production.
3. Write short notes on drilling riser.
4. Write short notes on top-tensioned riser.
5. Explain a steel catenary riser briefly.
6. Where are flexible risers deployed? What are their salient features?
7. Write a detailed note on the riser configurations that are commonly used in the offshore industry.
8. List various riser components and explain them briefly.
9. Explain the phenomenon of vortex-induced vibration in marine risers.
10. Explain the coupled motion of an offshore vessel and the riser. Why is it necessary to investigate this coupled motion?
11. What are the factors considered while choosing the material for marine risers?
12. Write short notes on steel, as a riser material.
13. Write short notes on titanium, as a riser material.
14. Write short notes on mooring lines and their configurations.
15. Explain the corrosion process, with a help of a basic corrosion cell.

CHAPTER 2

Functionally graded materials for marine risers*

2.1 Introduction

Marine risers are exposed to harsh weather conditions in the ocean environment, leading to a combination of unique requirements in terms of both the structural and functional perspectives. As the failure of marine risers can become highly catastrophic, leading to severe consequences, including significant environmental impacts, utmost care is taken even at the selection of the material for risers. In an offshore oil and gas-production system, several functions are carried out using a riser, namely, drilling, completion, production, export, or injection. As discussed in the previous chapter, marine risers are classified into free-standing, top-tensioned, catenary, and hybrid configurations (Chandrasekaran, 2015b, 2017; Chandrasekaran et al., 2015a,b). Marine risers are subjected to severe corrosion due to the presence of chlorides, hydrogen sulfides, and carbon dioxide gases. A marine riser is subjected to several environmental loads that arise from wave, current, ice, shock, and impact (Liu et al., 2016; Chandrasekaran, 2014; Chandrasekaran and Bhattacharyya, 2012).

Composite materials are also alternate materials for marine risers but failed to perform the intended function as a riser material. It is because they are highly prone to delamination under the combined action of both the mechanical and thermal loads. Further, delamination can result in a significant reduction in both the compressive and flexural strengths of the materials, which is undesirable for riser materials (Aslan and Daricik, 2016). The concept of a graded interlayer was proposed by Kawasaki and Watanabe (1987) for metal and ceramic bonding. It proved its superiority in terms of heat resistance and toughness to control crack propagation. However, under the high-temperature and high-pressure conditions, thermal stresses developed at the bonding interface causes delamination; thermal stresses occur due to the difference in the

* This chapter is co-authored by Dr. Hari Sreenivasan, Assistant Professor, School of Petroleum Technology, Pandit Deendayal Petroleum University, Gandhinagar. India, Email: hari.s@spt.pdpu.ac.in

Design of Marine Risers with Functionally Graded Materials
DOI: https://doi.org/10.1016/B978-0-12-823537-9.00004-9

coefficient of thermal expansion between the composition of materials. It further leads to the formation of cracks and propagates to cause severe deterioration in strength and durability. They also result in delamination at the interface of the materials. Unlike composites, functionally graded materials (FGMs) possess no distinct interface between the individual materials. Therefore, by the manufacturing process, it overcomes the damage posed due to delamination.

2.2 Functionally graded material

FGMs have found their successful applications in the defense, aerospace, and medical field; recent attempts are also made to find their practical use in the offshore industry (Albino et al., 2018). Besides, FGMs are also useful in fabricating cemented carbide cutting tools, thermal barriers, and antioxidation coatings. More frequent use of FGMs can also be found in the manufacturing of thermoelectric materials by grading the carrier concentrations (Koizumi, 1997). FGM can also be added to an existing component to improve its strength and durability. However, an appropriate surface finishing is to be carried out after deposition. FGM can also be used for various other applications—process industry where pipelines involving flow of chlorides and sulfides, pulp and paper industry where FGM clad plates are used for digesters, structural applications in coastal and offshore structures, and mechanical components of pump parts, which are susceptible to severe corrosion. One of the significant research challenges in manufacturing FGM is the functional grading of two metal components for a corrosion-resistant application using an additive manufacturing process. As such, the strength of the developed FGM shall depend significantly on the manufacturing process; one should ensure a reliable interface between the constituent materials to meet the strength of the application desired. The concept of FGM was introduced by Kawasaki and Watanabe (1987) for creating a thermal barrier. The cold metal transfer (CMT)-based WAAM is brought to commercial use in the year 2011. However, the application of the WAAM process to manufacturing FGM for a potential marine application is novel. CMT is a modified form of the gas metal-arc welding process, involving low current and voltage. It results in a significant reduction of the heat input in the deposition process. A dip transfer mode, which is digitally controlled, is deployed to transfer the molten droplet to the substrate.

In the manufacturing process of FGM, materials of desired characteristics are chosen, and their geometric compositions (not the metallurgical composition), in terms of thickness and number of layers, are varied continuously across the cross section. As shown in Fig. 2.1, FGMs possess a

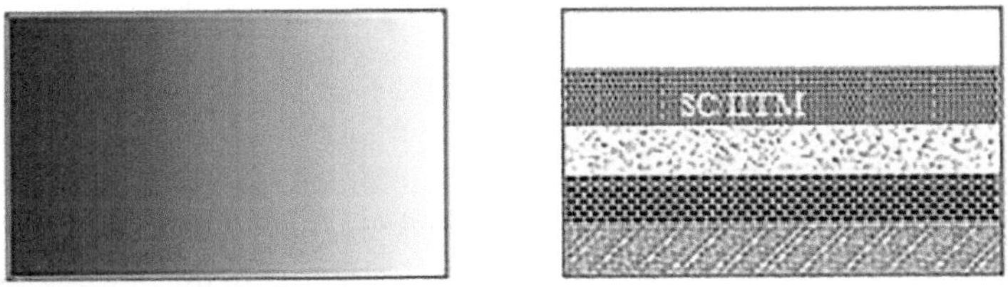

(A) (B)

Figure 2.1 Functionally graded material: (A) continuous variation and (B) stepped variation.

continuous variation (or) a step-wise grading of materials, depending on the application requirements. The composition and microstructure are changed along the cross section to generate a property gradient with the combined materials. Functional grading utilizes the advantages of mechanical, metallurgical, and structural characteristics of the original materials being graded. For example, duplex stainless steel and titanium show a higher corrosion resistance to hydrogen sulfide gases, chlorides, CO_2, and acidic environment. They also possess a higher resistance to stress corrosion cracking that is caused by chlorides in the marine environment (Chandrasekaran and Jain, 2016; Chandrasekaran and Gaurav, 2017).

Research studies showed that FGM could also have a composition of alloy and ceramics (Kan et al., 2018; Madec et al., 2018). However, metallic FGMs, using nickel-based superalloys, titanium, and stainless steel (Yeo et al., 1998; Gao and Wang, 2000; Rajan et al., 2010), are successful in various industrial applications. Further, they can be manufactured by additive manufacturing techniques. Alternatively, there are other conventional methods of manufacturing FGM, namely, centrifugal casting, spark plasma sintering, laser deposition, and powder metallurgy (Yuan et al., 2012; Shishkovsky et al., 2012; Jin et al., 2005; Übeyli et al., 2014; Chen et al., 2017; Bermingham et al., 2015). The realization of the parts of a complex alloy became possible with the advancements in material science and advanced manufacturing techniques. As a result, several additive manufacturing techniques for manufacturing FGM, namely, selective laser melting, direct laser melting process, electron beam melting, and wire arc additive manufacturing (WAAM), are also in practice. The author wishes to present the details of one of the manufacturing methods used to fabricate FGM in the lab-scale: WAAM. It is one of the advanced techniques for manufacturing metallic components for engineering applications.

2.3 Wire arc additive manufacturing

The WAAM is a compatible manufacturing process as it enables the metallurgical composition of materials to be user-defined. The step-wise addition of materials in layers is practiced in the WAAM process to fabricate a three-dimensional component. The manufacturing process is guided by a digital model in which chosen materials are functionally graded; material in the form of wires is interchanged during the process of deposition. WAAM overcomes one of the severe limitations of the powder-bed process of manufacturing in terms of the deposition rate, equipment cost, and the cost of raw material. In the WAAM process of manufacturing the metallic powder is replaced by wires, while an electric arc is used as the heat source. Therefore the WAAM manufacturing process is considered to be economical and highly versatile as it is capable of manufacturing more large sizes of any geometric shapes. It is because the higher deposition rate, components of larger size, can be manufactured in a shorter time with a minimum material wastage (Williams et al., 2016; Martina et al., 2012). It also finds use in the repair of metallic components to restore the component to its full density and thereby reduces the processing time. However, the dimensional accuracy of the fabricated part depends on the precision of the motion system used for the guidance of the welding torch. Then machine and material cost in WAAM is lesser in comparison to other additive manufacturing techniques (Ma et al., 2015; Shen et al., 2016a,b). In the recent applications, it is shown that the speed ratio of the wire feed is altered with a twin-wire feeding arrangement to exercise a desirable control on the deposition rate and the composition.

High-pulse current is supplied to form an arc between the electrode wires and the substrate, resulting in the melting of the filler tip of the advancing electrode. A sharp drop in voltage indicates a short circuit, after which the current is lowered, and the filler wire is retracted. This retraction causes the detachment of the molten droplet to the substrate, which is further assisted by the surface tension and Lorentz force in the weld pool. The metallic wire is advanced, mechanically using a secondary wire-feeder, which operates at a higher frequency. It is integrated into the torch, which aids the droplet detachment. Transfer of the droplet occurs at an instance when the current is lower, and the arc is almost absent. It is due to this process in WAAM that the name of CMT is associated with WAAM. The CMT was used initially used to join aluminum, which needs a lower amount of heat input (Zhang et al., 2009). Arc burning time during the manufacturing process is set to the minimum such that the heat generated is just enough to melt the wire and fuse it to the base metal.

2.4 Functionally graded material to the marine riser

Carbon–manganese steel of select grades is one of the most commonly used materials for manufacturing marine risers (Det Norske Veritas, 2012). To improve its performance a corrosion-resistant alloy of duplex stainless steel is functionally graded with the carbon–manganese steel along the inner portion of the riser. Duplex stainless-steel possesses a higher corrosion resistance toward hydrogen sulfide gases, carbon dioxide, and acidic conditions (Det Norske Veritas, 2018). It is also capable of resisting the formation of cracks by delaying the crack initialization, which otherwise forms rapidly in the presence of chlorides. The chosen graded composition also possesses a higher strength, which is desirable in the deepwater environment. Both the materials, chosen for manufacturing FGM, individually are widely acceptable for marine usage and are recommended in the Det Norske Veritas codes (see Table 2.1 for more details). In the WAAM process, individual wire spools of ER2209 and ER70S6 are used for successive deposition of duplex stainless steel and X52 carbon–manganese steel, respectively. The stress–strain curve of X52 carbon–manganese steel under axial tension is shown in Fig. 2.2 (Lyapin et al., 2015).

As a part of the real-time implementation of FGM in marine risers, the riser configuration in terms of cross-sectional thickness and diameter is considered to be similar to that of the riser used in the Auger tension leg platform (TLP) (Phifer et al., 1994). The steel catenary riser, installed in Auger TLP, is considered a reference riser as no FGM risers are installed in the offshore industry, so far. Riser, deployed in the Auger field, has a wall thickness of 0.688″ (17.47 mm) and an outer diameter of 12.7″ (323.85 mm). However, in the present discussion, the author attempts to examine the applicability of FGM riser with similar geometric properties.

The marine riser, considered in the present case, is fabricated using FGM. It comprises 22Cr duplex stainless steel along its inner surface while

Table 2.1 Materials for riser and code applicability.

Material	Code compliance
Duplex stainless steel (22Cr) ER2209	DNVGL-RP-F112
Carbon–manganese steel ER70S6	DNV-OS-F101

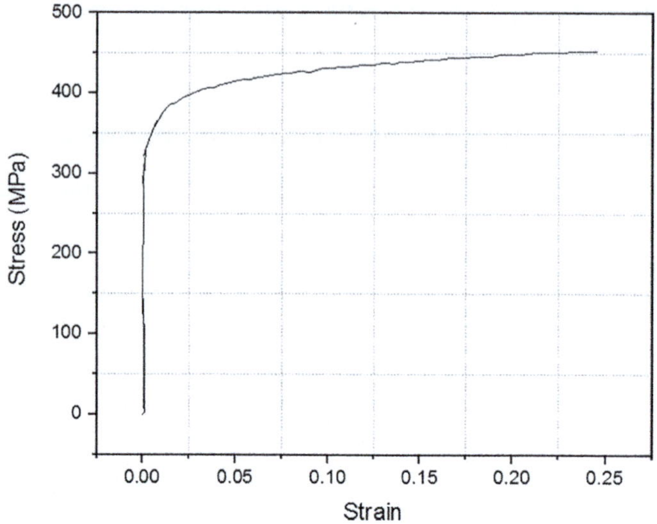

Figure 2.2 Stress—strain curve of X52 carbon—manganese steel.

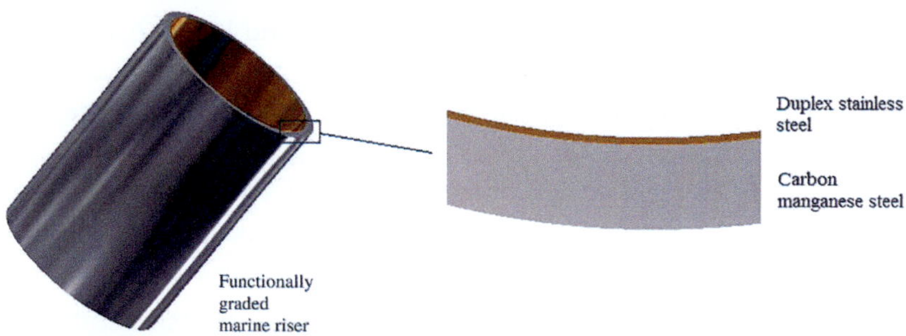

Figure 2.3 Section of the proposed FGM riser. *FGM*, Functionally graded material.

maintaining the same thickness as that of the reference riser. The resulting FGM riser is shown in Fig. 2.3, while its composition of materials is shown in Table 2.2.

The wall thickness of the FGM riser is assessed for the code requirements and found to be satisfied (Chakrabarti, 2005). The desired wall thickness of a marine riser is given by the following expression, after accounting for the corrosion allowance:

$$t_{\min} = \frac{P_i D_{\text{nom}}}{2\,\eta_h \sigma_y} + t_{\text{cor}} \tag{2.1}$$

Table 2.2 Functionally graded marine riser.

Layer	Material	Thickness (mm)
1	Duplex stainless steel (22Cr) 2209	3
2	Carbon–manganese steel ER70S6	14.47
	Total thickness	17.47

where σ_y, D_{nom}, and P_i are the yield strength, nominal diameter, and the internal pressure, respectively. For riser applications the usage factor is (η_h) is 0.6, resulting in a corrosion allowance (t_{cor}) of 6.265 mm. Besides, as per international design practice, a minimum internal corrosion allowance of 3 mm is to be provided for carbon–manganese steel risers (Det Norske Veritas, 2012). Along the inner surface of the riser a corrosion-resistant liner of 3 mm thickness was installed to handle scouring and CO_2 requirements (Gouveia et al., 2015).

2.5 Materials and manufacturing

A stainless-steel substrate of size $330 \times 30 \times 6\ mm^3$ is used to deposit the materials for WAAM. The power source used for the deposition process is Fronius CMT TransPuls Synergic 4000, while the welding torch used for the deposition of the materials is integrated to a triaxial computer numeric control (CNC) machine, as shown in the Fig. 2.4. G-codes can be used to obtain the desired motion of the welding torch in X, Y, and Z coordinates. The weaving pattern of deposition is used to obtain the desired layers of deposition. G-codes are programmed to obtain the desired weaving pattern, as shown in Fig. 2.5A; bidirectional deposition strategy is used to enable the deposition of multiple layers, as shown in Fig. 2.5B. Bidirectional deposition enables the disruption of columnar grains during the process of solidification (DebRoy et al., 2018).

Filler wires of 1.2 mm in diameter are used in the manufacturing process, while the contact tip to work distance (CTWD) is set to 15 mm. Mechanical clamps are used to restrain the substrate on the base plate. Before proceeding for the deposition, appropriate weld-deposition parameters, namely, current, weld-speed, and the rate of the wire-feed, are optimized. The desired, functionally graded build is obtained in stages. During the process of deposition, weaving width is considered as half of that of the optimized transverse bead width, which is obtained in a straight weld. As explained earlier, the functionally graded build of two materials is

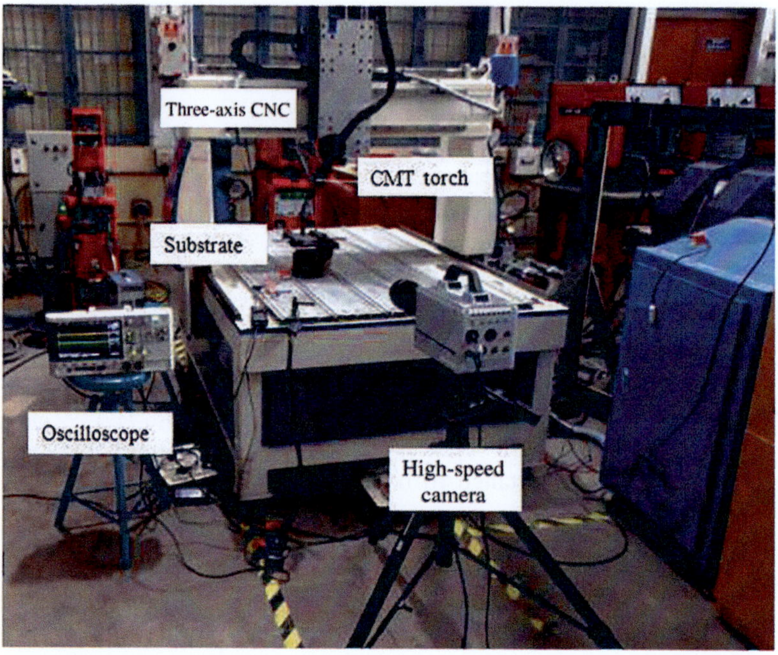

Figure 2.4 Cold metal transfer Gas Metal Arc Welding Torch integrated to CNC machine. *CNC*, Computer Numeric Control.

manufactured in two stages. The chemical composition of the filler material, ER70S6, which is used for the deposition of carbon–manganese steel and ER2209, used for the deposition of duplex stainless steel is shown in Table 2.3.

The arc energy (or) the heat input per unit length required for the deposition process is given by:

$$\mathrm{IP} = \frac{1}{n}\sum_{i=0}^{n} U_i I_i \tag{2.2}$$

where U_i is the instantaneous voltage, I_i is the instantaneous current, and IP is the instantaneous power. Arc energy per unit length is given by:

$$\text{Arc energy} = \frac{(\eta \cdot \mathrm{IP})}{v} \tag{2.3}$$

where η is the efficiency as 0.8, and v is the deposition speed, used in the welding process. The deposition parameters optimized for carbon–manganese steel and duplex stainless steel are shown in Table 2.4.

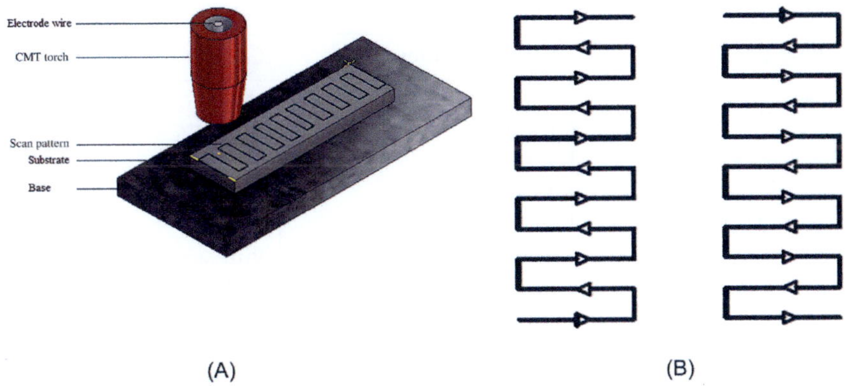

Figure 2.5 Additive manufacturing of FGM: (A) scan weaving path and (B) bidirectional weaving for successive depositions. *FGM*, Functionally graded material.

Table 2.3 Chemical composition of the filler material.

Material	By weight (%)	
	Filler material ER70S-6	Filler material ER2209
Carbon (C)	0.06	0.008
Manganese (Mn)	1.45	1.62
Silicon (Si)	0.90	0.49
Phosphor (P)	0.02	0.016
Sulfur (S)	0.03	<0.0005
Nickel (Ni)	0.12	8.59
Chromium (Cr)	0.12	22.93
Copper (Cu)	0.4	0.083
Molybdenum (Mo)	< 0.15	3.02
Nitrogen (N)	NA	0.152
Vanadium (V)	0.03	NA
Cobalt (Co)	NA	0.041
Ferrum (Fe)	Remainder	Remainder

The voltage–current waveforms that are used as an input for the process of deposition are recorded using a 75 MHz, two-channel InfiniiVision Digital Storage Oscilloscope.

The recorded waveforms are plotted in Figs. 2.6 and 2.7, for carbon–manganese steel and duplex stainless steel, respectively. Three layers of duplex stainless steel are deposited on the stainless-steel substrate using the optimized welding parameters, as shown in Fig. 2.8. Five layers of

Table 2.4 Deposition parameters for functionally graded material composition.

Welding parameters	Duplex stainless steel	Carbon–manganese steel
Welding process	CMT	CMT
Filler	ER2209	ER70S6
Filler diameter	1.2 mm	1.2 mm
Mean current	119 A	107 A
Mean voltage	12.86 V	16 V
Shielding gas	Pure argon	Pure argon
Shielding gas flow rate	16 L/min	16 L/min
Arc energy	222.59 kJ/m	256.8 kJ/m
CTWD	15 mm	15 mm
Welding speed	330 mm/min	320 mm/min

CMT, Cold metal transfer; *CTWD*, contact tip to work distance.

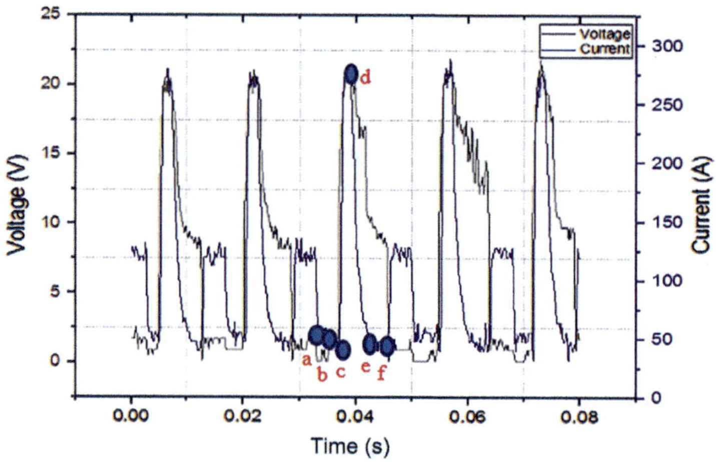

Figure 2.6 Input voltage–current waveform for ER70S6 carbon–manganese steel deposition.

carbon–manganese steel, using the optimized parameters, are deposited on the previously deposited duplex stainless steel to form a FGM block of size $300 \times 25 \times 45$ mm^3 (see Fig. 2.8). Fig. 2.9 shows the deposition of duplex stainless steel on the substrate and the successive deposition of carbon–manganese steel on the layer of duplex stainless steel.

2.6 Mechanical characteristics

A high-speed video camera with 6000 fps is deployed to capture the metal-transfer characteristics during the deposition process. Images captured at

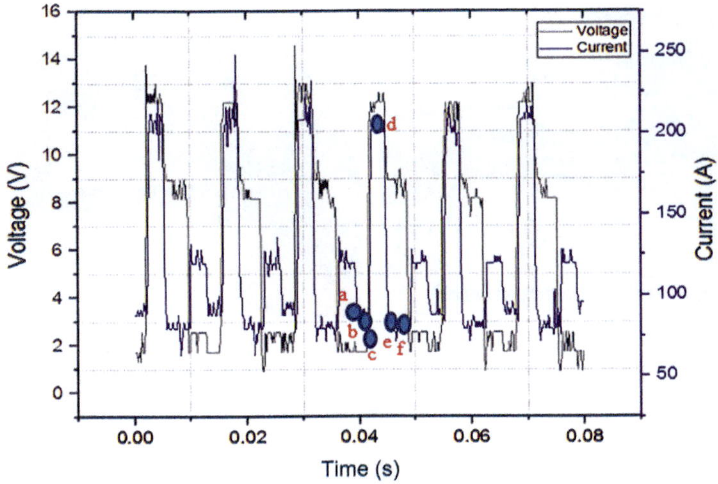

Figure 2.7 Input voltage–current waveforms for ER2209 duplex stainless-steel deposition.

Figure 2.8 Functionally graded combination of duplex stainless steel and carbon–manganese steel.

various time intervals of the significant events of deposition during a single cycle are compared with that of the current–voltage waveform and subsequently analyzed using an oscilloscope. X-ray computed tomography is also carried out in three different configurations, namely, top, front, and side orientations using General Electric Phoenix v |tome| x-s to assess the quality of the build. Test specimens of 5 mm thickness are cut, which represent samples from the axial direction of the marine riser, as shown in Fig. 2.10. The samples are cut from the build after undergoing operations such as cutting, milling, and wire electro discharge machining (EDM). The test

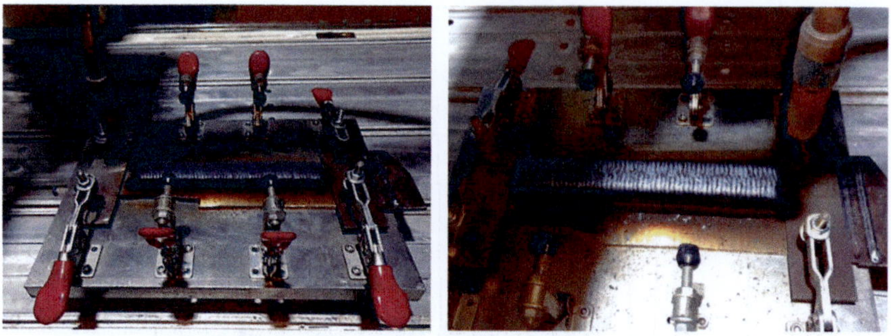

(A) (B)

Figure 2.9 (A) Deposition of duplex stainless steel on the substrate and (B) deposition of carbon–manganese steel on duplex stainless steel forming an FGM. *FGM*, Functionally graded material.

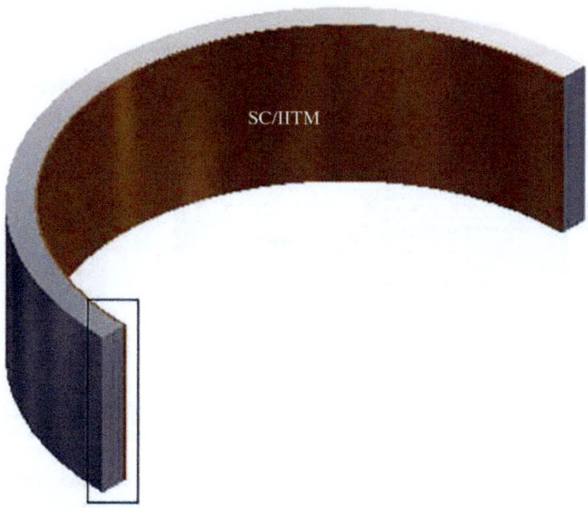

Figure 2.10 A sectional view of FGM material along the longitudinal axis of the riser. *FGM*, Functionally graded material.

specimens are adhering to the ASTM E8 dimensions, as shown in Fig. 2.11. A gauge width of 17.47 mm is maintained in the test specimen, as shown in Fig. 2.12, comprising 14.47 mm of carbon–manganese steel and 3 mm duplex stainless steel (refer to Table 2.2 for details).

Poisson's ratio is estimated by measuring the longitudinal tensile strain and the transverse compressive strain. Zwick Roell Z100 universal testing machine, which is integrated with a video extensometer is used for the

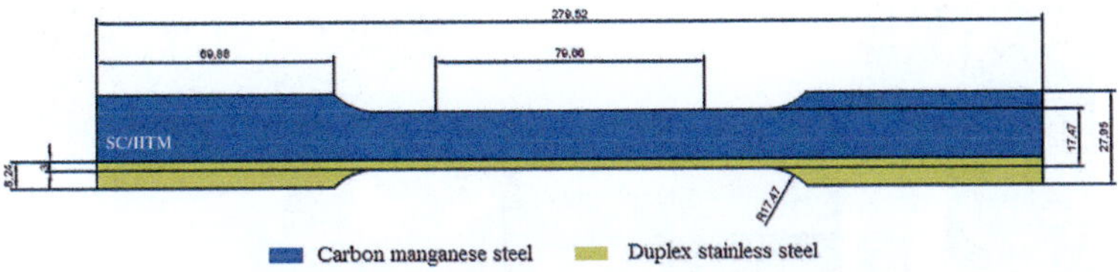

Figure 2.11 Dimensions of the test specimen to be cut out from FGM build. *FGM*, Functionally graded material.

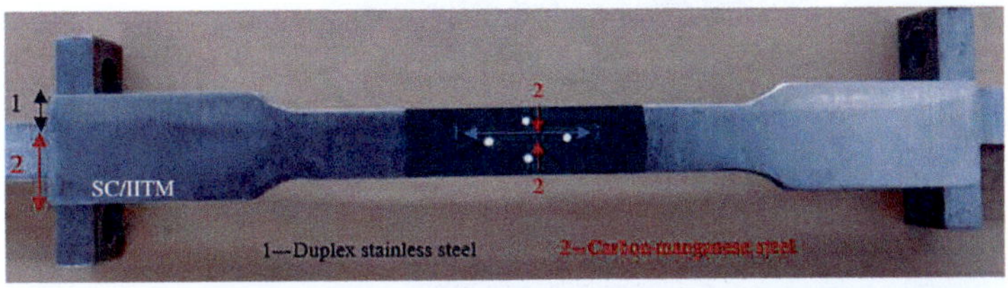

Figure 2.12 Optical markers for measuring the longitudinal tensile strain (1-1) and transverse compressive strain (2-2).

experimental investigations. Fig. 2.13 shows the details of the experimental setup. Four white marks, as seen in Fig. 2.12, are the reference points located on the surface of the FGM build, which serves as an optical marker to measure the strain. The blue line, marked along with the two white dots, is used to calculate the tensile strain in the longitudinal direction. The two white dots, marked vertically, are used to determine the transverse strain, which is compressive. Once the test specimen is optically marked, the specimens are tested at a strain rate of $1 \times 10^{-3}\ s^{-1}$.

An ASTM E8 subsize dimensions are cut out from the build, as shown in Fig. 2.14. The three test specimens cut from the build are shown in Fig. 2.15. Universal testing machine Instron 3367 is used for the tensile test of three FGM samples at a strain rate of $1 \times 10^{-3}\ s^{-1}$. A gauge length of 10 mm is maintained, which comprises equal parts of carbon–manganese steel and duplex stainless steel.

EDM is used to cut the samples for microstructure analysis. A section is cut transverse to the build comprising the interface between the carbon–manganese steel and duplex stainless steel. The test section, thus, obtained is polished using an emery paper of grades 220–2000 one after the other. For preparing the sample for metallographic investigations, the surface finish is further improved by polishing the samples with 6–12, 3–4, and 0.5–1 μm diamond pastes. The etchant used for carbon–manganese steel is 4% Nital; electrolytic etching is done on the duplex stainless steel using 10% NaOH.

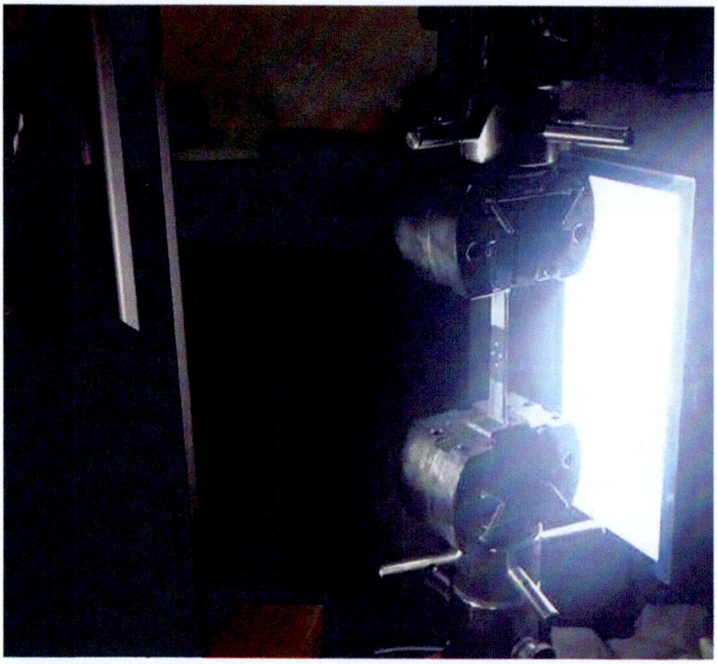

Figure 2.13 Test specimen attached to universal testing machine with a video extensometer.

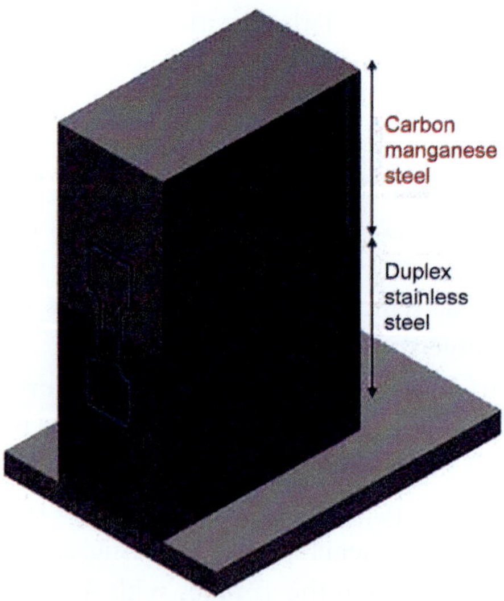

Figure 2.14 Illustration showing the region where the test specimen was cut out from the functionally graded build.

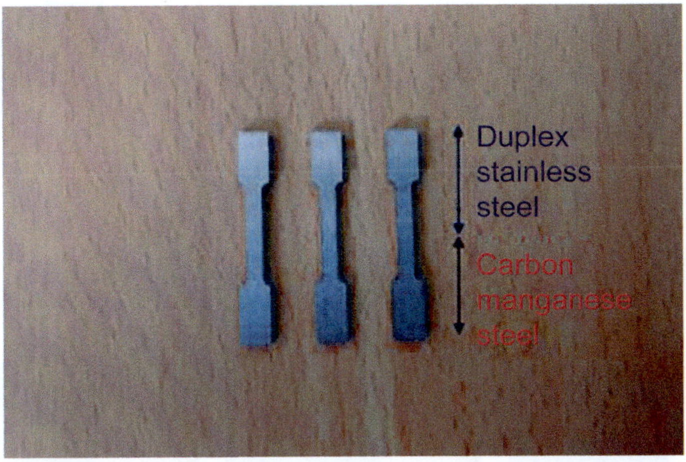

Figure 2.15 Test specimen dimensions to be cut out from the functionally graded build (all dimension in mm).

The microstructure in the required magnification is obtained using an Optical Microscope—LEICA DFC295.

To determine the elemental composition across the interface, and energy dispersive X-ray (EDX) analysis is carried out using Quanta 200F scanning electron microscope (SEM). Observations are recorded for a 35 μm length, measured from the face of the carbon–manganese steel to that of the duplex steel. The microhardness test is also carried out on the polished section, which is extracted from the FGM build. The section is examined for microhardness from the face of duplex stainless steel to that of carbon–manganese steel layers. The hardness values are obtained along the centerline of the build at a spacing of 1 mm c/c between the corresponding points. A net force of 200 g is applied for a dwell time of about 10 s using a Wolpert Wilson micro Vickers digital hardness tester. The samples obtained after tensile testing are subjected to fractography analyses to determine the type of failure. X-ray diffraction (XRD) studies of the FGM interface is carried out using Brucker D8 XRD equipment.

2.7 Metal-transfer characteristics

High-speed video analysis is performed to assess the metal-transfer characteristics of the FGM build. The waveform, recorded in the oscilloscope, is compared with the frames captured using a high-speed camera for a single cycle of deposition. High-speed imaging of the metal-transfer of the filler

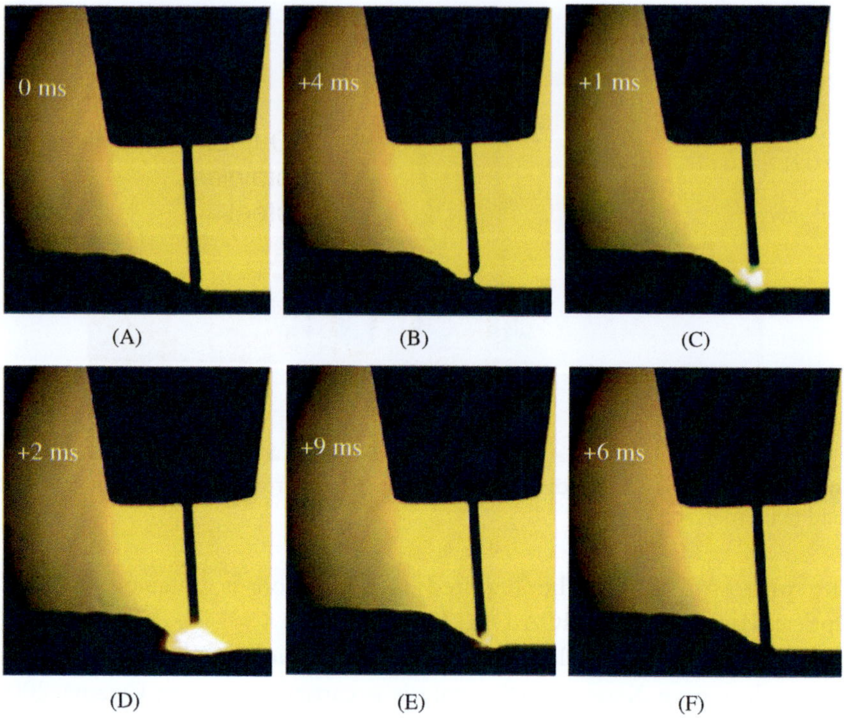

Figure 2.16 Metal-transfer characteristics of carbon–manganese steel.

wire of the carbon–manganese steel in the CMT mode is shown in Fig. 2.16. The cycle of commencement occurs at the point of commencement of the short circuit. It is marked as "a" in the waveform. After the duration of 4 ms the filler wire retracts and leads to the detachment of the molten metal, which is marked as "b." It results in the creation of wavy disturbances in the molten pool, which is marked as "c." Upon detachment, the current rises and results in the arc formation. The further current increases and reaches its maximum value when the arc is fully formed, which is marked as "d." Subsequently, the arc gradually extinguishes as the wire advances toward the pool, as seen in "e." Again, short-circuiting occurs, which is marked as "f," enabling the cycle to repeat. Fig. 2.17 shows the metal-transfer characteristics of duplex stainless steel.

2.8 X-ray computed tomography

The FGM build, manufactured for a reference marine riser, is also experimentally investigated in all the three orientations to assess porosity and the

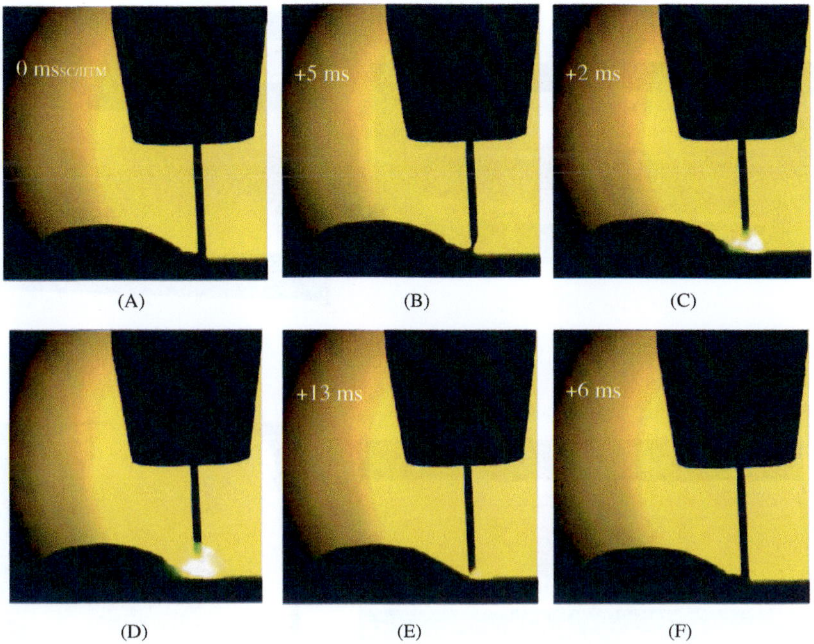

Figure 2.17 Metal-transfer characteristics of duplex stainless steel.

formation of microcracks, if any, during the manufacturing process. The results obtained from the computed tomography are shown in Fig. 2.18. It is seen from the figures that the FGM build is free from both porosity and microcracks. The carbon–manganese steel and duplex stainless-steel interface are also free from any such defects. The analysis also confirms that the WAAM process, used to fabricate the FGM build for the marine riser, is suitable to manufacture excellent material, which can be used for engineering applications.

2.9 Tension test

FGM samples are further investigated under axial tension at a strain rate of $1 \times 10^{-3}\,s^{-1}$. The stress–strain curve of the FGM sample is shown in Fig. 2.19; it corresponds to the longitudinal strain, while Fig. 2.20 shows the stress–strain curve in the transverse direction. It is seen from the figures that the ultimate strength is about 602 MPa with an yield strength of about 403 MPa, while the modulus of elasticity is computed as 213 GPa. Fig. 2.21 also shows a comparison of the stress–strain curves of

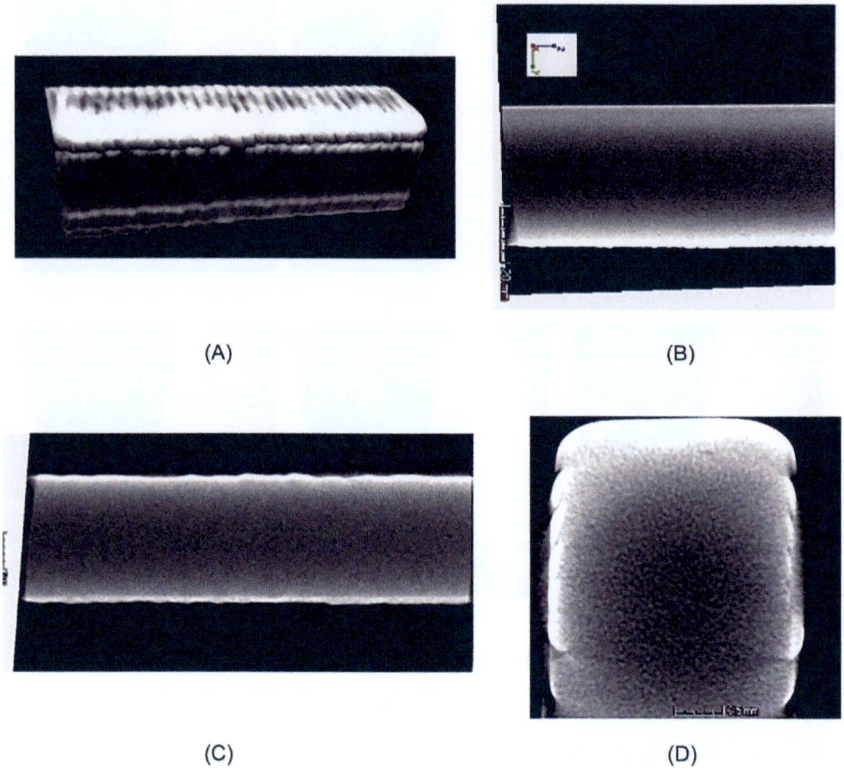

Figure 2.18 X-ray computed tomography carried out on the FGM build: (A) 3D, (B) top, (C) side, (D) front. *FGM*, Functionally graded material.

X52 steel used for the marine risers and that of the FGM mould, proposed for the marine riser. As seen, both the strength and ductility characteristics of the FGM are found to be better than X52 steel.

The fractured specimen, shown in Fig. 2.22, highlights the failure as the optical markers are moved apart. From the failure surface, it is inferred that there is no failure at the interface of the two-graded materials. Table 2.5 shows the comparison of the interface strength. It is observed that the yield strength and percentage elongation of FGM is closer to that of the X52 steel, while the ultimate strength is higher by about 6%. The fractured FGM sample after the tension test also confirms that the failure has occurred in the carbon–manganese steel portion, while the interface remains intact. A comparison of the interface strength of FGM samples with that of X52 material is shown in Table 2.6.

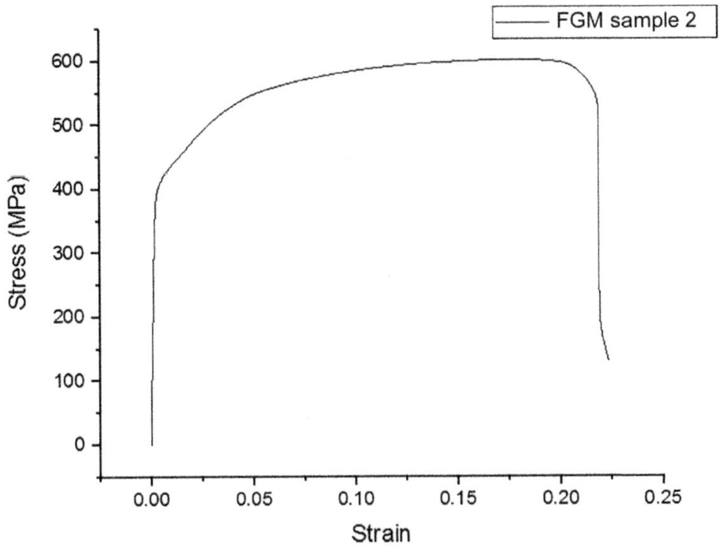

Figure 2.19 Stress–strain curve of FGM sample-2 along the longitudinal (tensile) direction. *FGM*, Functionally graded material.

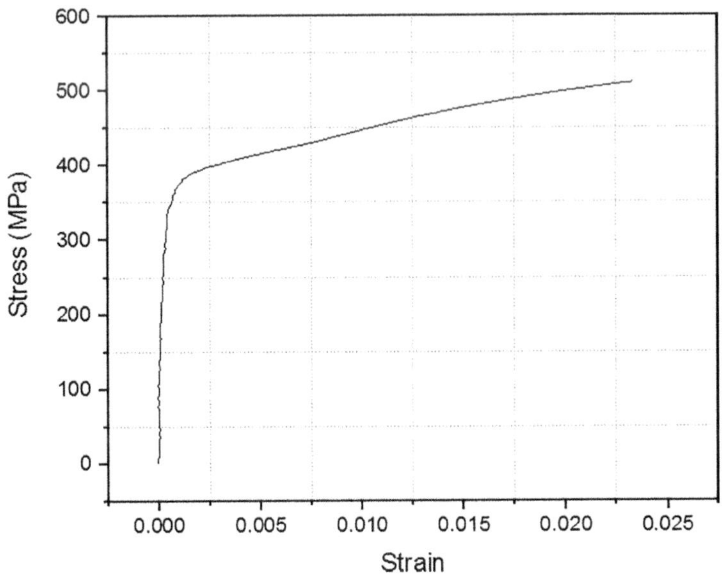

Figure 2.20 Stress–strain curve of FGM sample-2 along the transverse direction. *FGM*, Functionally graded material.

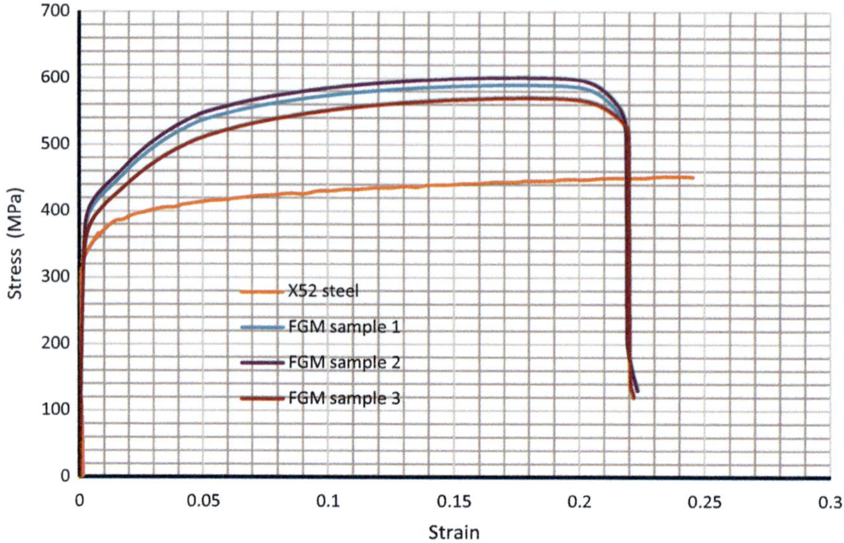

Figure 2.21 Stress–strain curves of FGM samples and X52 steel (longitudinal direction). *FGM*, Functionally graded material.

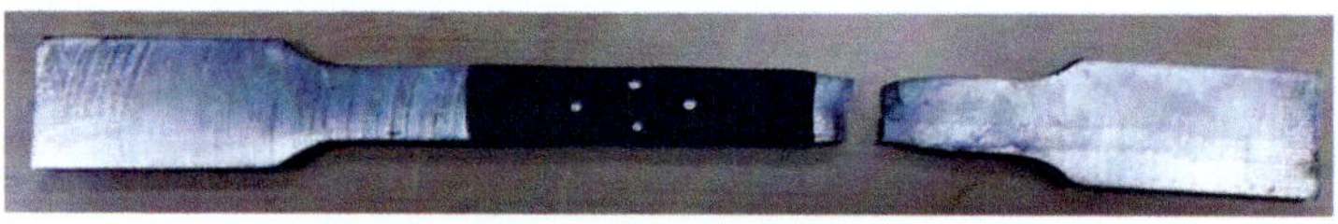

Figure 2.22 Fractured FGM sample during the tension test. *FGM*, Functionally graded material.

Table 2.5 Comparison of mechanical properties of functionally graded material (FGM) and X52 marine steel.

Material/parameters	X52 marine riser steel (Lyapin et al., 2015)	FGM
Young's modulus (GPA)	210	209.66 ± 4.48
Yield strength (MPa)	358	390.66 ± 12.23
Ultimate strength (MPa)	453	587.66 ± 12.76
Strength ratio	1.265	1.50 ± 0.02
Ductility ratio	32.207	45.47 ± 0.82
Tensile toughness (J/m^3)	104.92	120.50 ± 2.84
Poisson's ratio	0.3	0.30 ± 0.07
% Elongation	20	22.31 ± 0.11

Table 2.6 Comparison of the interface strength of functionally graded material (FGM) with X52 marine riser steel (ASTM E8 subsize specimen).

Material	Yield strength (0.2% proof stress) (MPa)	Ultimate strength (MPa)	% Elongation
FGM	340	483 ± 2.5	16.02 ± 0.4
X52 (Lyapin et al., 2015)	358	455.05	21

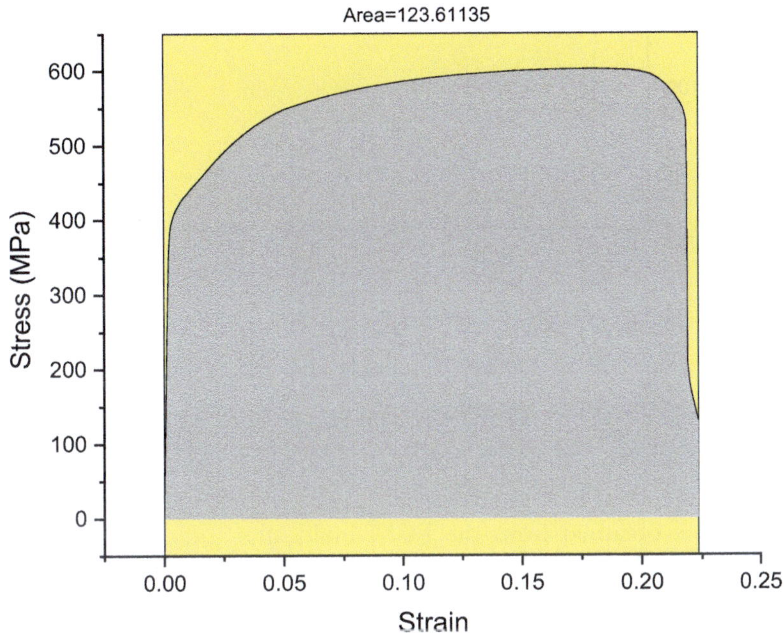

Figure 2.23 Toughness of FGM sample-2. *FGM*, Functionally graded material.

2.10 Structural properties

Structural properties of the functionally graded combination, proposed for the marine riser is obtained from the stress–strain curves, in tension. The tensile toughness or deformation energy of the FGM sample-2 and X52 marine riser steel are compared. While the former is measured as 123.611 J/m^3, the latter is only 104.924. Figs. 2.23 and 2.24 show the toughness measurements of FGM and X52 steel, respectively.

It is observed from the table that the FGM mould possesses superior mechanical properties in comparison to that of X52 marine riser steel. It is important to note that the tests are carried out without any postprocessing operations; mechanical properties are evaluated in the built condition.

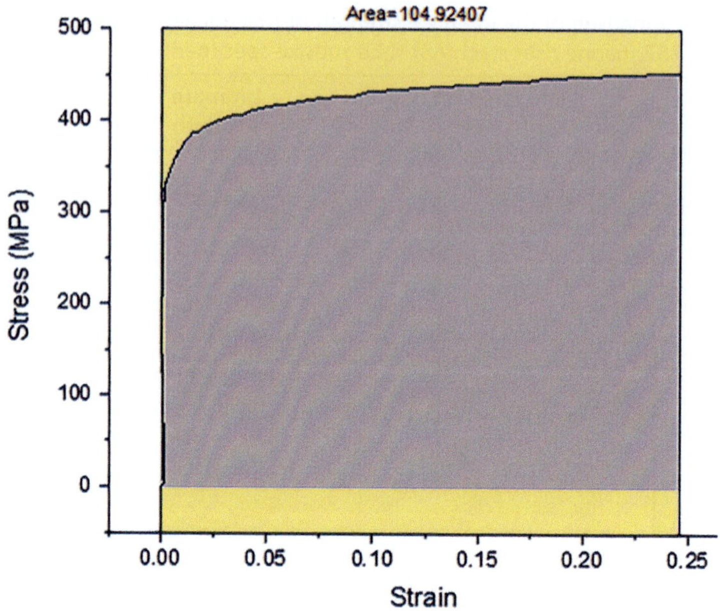

Figure 2.24 Toughness of X52 steel.

2.11 Microstructure analyses

The section cut from the FGM mould, intended for using a new material for the marine riser, is now examined for the microstructure analysis. The test section is obtained from the FGM build after grinding it to a high-surface finish. The test section consists of layers of duplex stainless steel, interface, and the layers of carbon—manganese steel. The study is carried out at seven different sections, to arrive at a detailed conclusion on the metallic microstructure of the new FGM mould. Fig. 2.25 shows the details of test sections where the microstructure analyses are carried out. As seen from the figure, test sections 1—5 are located in the carbon—manganese steel face, while section 6 is located on the interface between the carbon—manganese steel and the duplex stainless steel; section 7 is taken on the duplex stainless-steel face.

The microstructure at location 1 is shown in Fig. 2.26. An equiaxed grain formation is observed with the microstructure comprising pearlite in a matrix of ferrite. The microstructure at location 2, as seen in Fig. 2.27, shows an equiaxed formation of pearlite in a matrix of ferrite but also contains traces of martensite. At location 3, as seen in Fig. 2.28, a higher concentration of martensite is observed. The microstructure at location 4, as

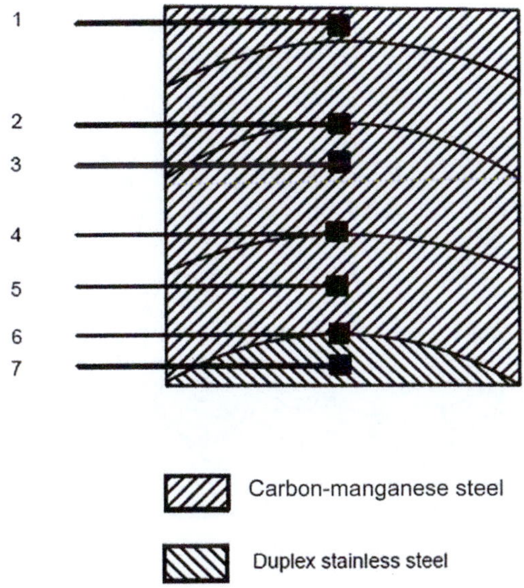

Figure 2.25 Sections for microstructure analyses.

Figure 2.26 Microstructure of carbon–manganese steel at location 1.

seen in Fig. 2.29, shows a clear distinction between the martensite and the pearlite formation in a matrix of ferrite. The microstructure at location 5 is shown in Fig. 2.30. It is seen that the martensite formation is observed in the face of the carbon–manganese steel. Fig. 2.31 shows the microstructure

Figure 2.27 Microstructure of carbon–manganese steel at location 2.

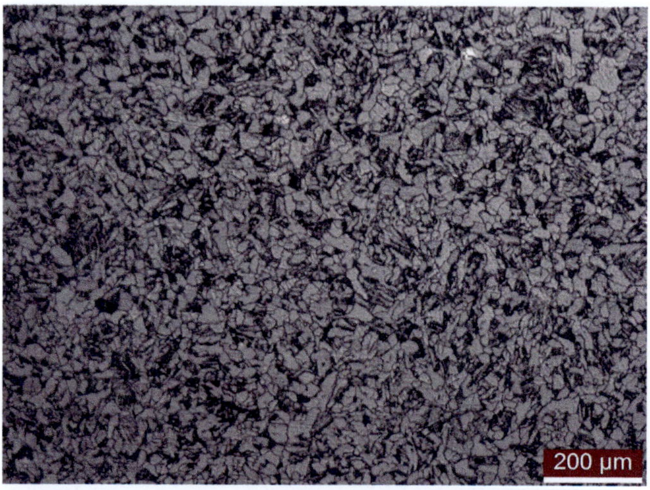

Figure 2.28 Microstructure of carbon–manganese steel at location 3.

at location 6. It is the location where there is an interface between the duplex stainless steel and carbon–manganese steel. It is seen from the figure that there is a complete fusion between the individual materials with no detection of any defects. At location 7, as seen in Fig. 2.32, microstructure shows austenitic (bright) and ferritic (dark) grains, which is an indication of a total solidification of the duplex stainless steel.

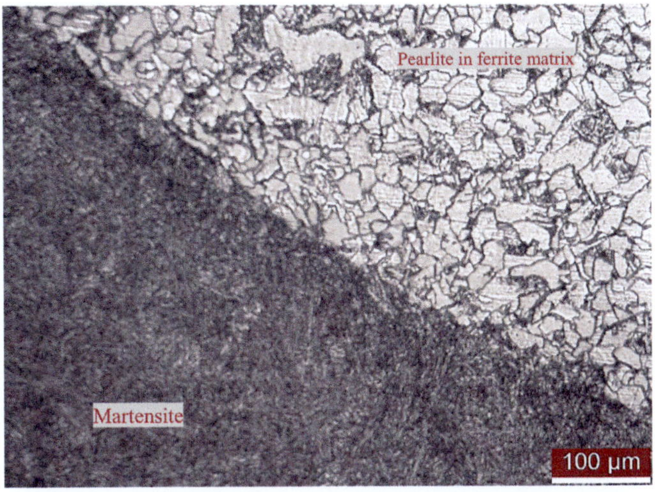

Figure 2.29 Microstructure of carbon–manganese steel at location 4.

Figure 2.30 Microstructure of carbon–manganese steel at location 5.

2.12 Energy dispersive X-ray analysis

At the interface of carbon–manganese steel and duplex stainless steel, an elemental line EDX scan analysis is carried out to determine the chromium variation across the interface. The image of the X-ray analysis is shown in Fig. 2.33. As seen in the figure, the chromium content in duplex stainless steel is around 20%–22% at 32–35 μm from the interface. It is closer to the chromium content in the 22Cr Duplex stainless steel used in the

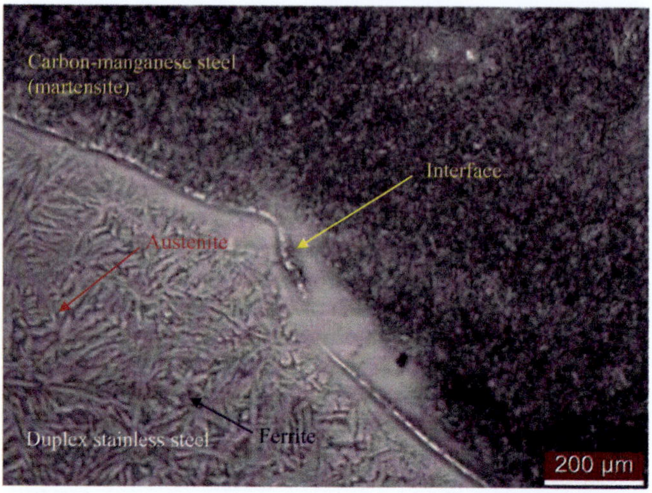

Figure 2.31 Microstructure of carbon–manganese steel at location 6.

deposition process (refer to Table 2.3). It is important to note that the corrosion-resistance behavior of duplex stainless steel is mainly due to the presence of chromium. Fig. 2.34 shows the variation of the chromium, molybdenum, and nickel in the EDX test carried out across the interface region. The EDX analysis carried out on the FGM mould confirms that the 3 mm layer of duplex stainless steel, chosen for the FGM as riser material, is sufficient to provide the corrosion resistance to the desirable level. It is also observed that there is no substantial enrichment of chromium at the interface toward the carbon–manganese steel.

2.13 Microhardness

The variation in microhardness along the build is shown in Fig. 2.35. It is observed that in the region of duplex stainless steel, the Vickers hardness varies between 267 and 235, and in the carbon–manganese steel, it is between 133 and 197. More interestingly, at the interface of 2 mm thick, the hardness values are observed to be in the range of 307–320. Near the vicinity of the interface zone, the martensite formation increases hardness. This increase in hardness contributes to the overall increase in the strength of the material. The region of duplex stainless steel is of a considerably higher hardness when compared to carbon–manganese steel, which provides additional strength to the FGM material.

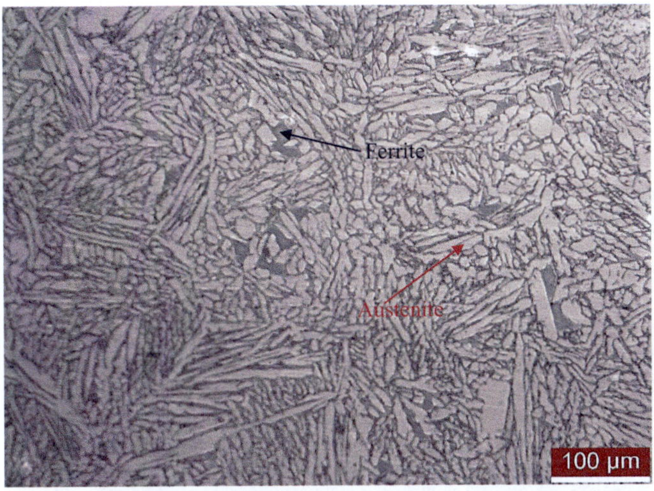

Figure 2.32 Microstructure of duplex stainless steel at location 7.

2.14 Fractography

The fractured zone of the FGM mould, during the experimental investigations, is further analyzed by extracting a section comprising both carbon–manganese steel and duplex stainless steel along the tensile sample cross section. Fractography study is carried out at both these locations of the tensile sample, as shown in Fig. 2.36. It is seen from the images that the dimples are prominent and of a larger diameter in carbon–manganese steel when compared to that of duplex stainless steel. A few smooth cleavages present in the duplex stainless-steel side can also be noticed, indicating that the carbon–manganese steel has contributed to the improvement of ductility in the FGM mould. It is also observed that both duplex stainless steel and carbon–manganese steel show a ductile failure. Fig. 2.37 shows the fractography images obtained at the failed portion of the carbon–manganese steel of the ASTM E8 subsize specimen. Under the application of load, there occurs a stress-partitioning between the two materials of the FGM mould.

2.15 Pitting corrosion

A square-shaped sample is cut from the build of carbon–manganese steel and duplex stainless steel. They are subjected to pitting corrosion test as per ASTM G48-11 method A. The test solution used is 100 g of reagent

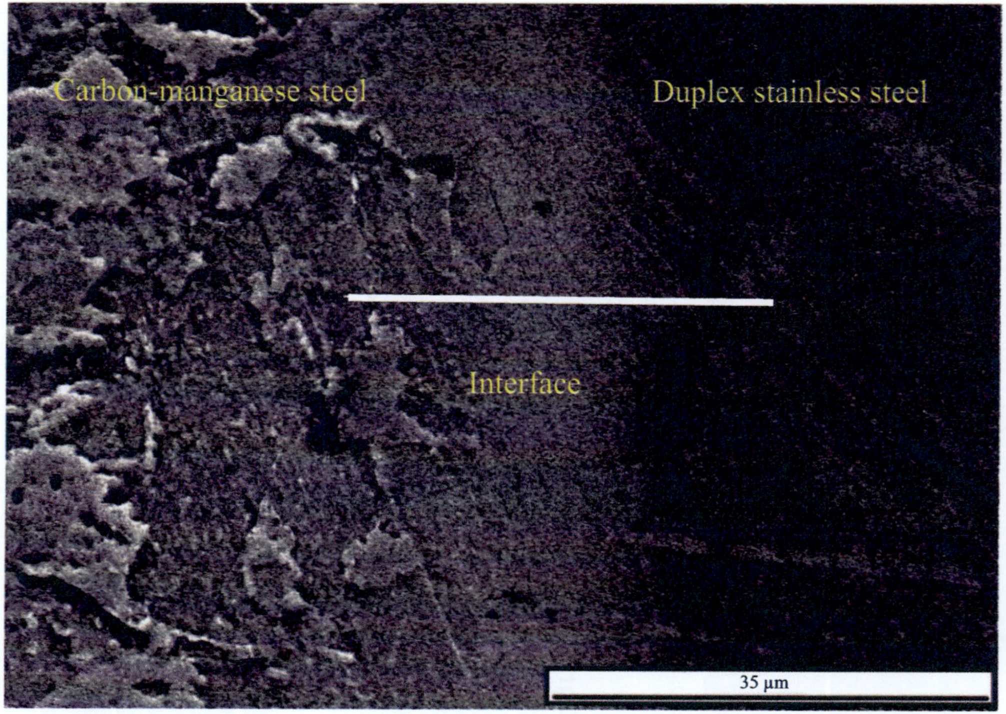

Figure 2.33 EDX test at the interface of FGM mould. *FGM*, Functionally graded material.

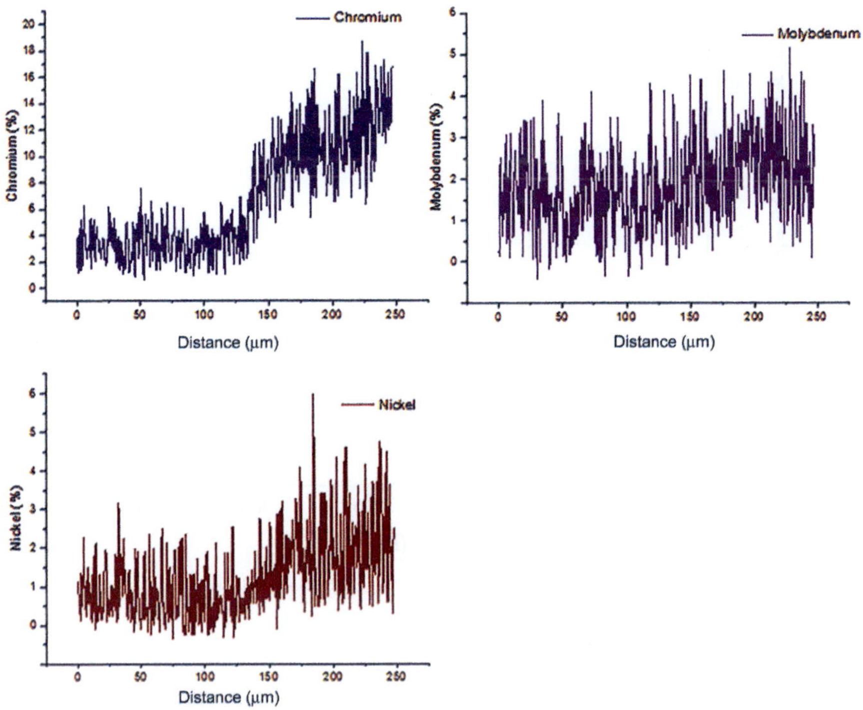

Figure 2.34 Variation of the chromium, molybdenum, and nickel in the Line EDX test carried out across the interface region.

grade ferric chloride in 900 mL of distilled water. All the surfaces of the specimens were ground finished using 120-grade emery paper. The accelerated corrosion test duration is 24 h. The corrosion rate is determined by a change in weight and surface area. The corrosion rate obtained for duplex stainless steel is 70.77 g/cm^2, and that of carbon–manganese steel, which corresponds to X52 steel (marine riser) is 823.35 g/cm^2.

2.16 X-ray diffraction analysis

It is observed from the XRD output shown in Fig. 2.38 that there is a central peak $Cr_{0.1}$ and $Fe_{1.9}$ (Cr_2O_3) indicating the presence of chromium oxide at the interface regions of the FGM. The formation of chromium oxide gives duplex stainless steel its corrosion resistance.

Detailed investigations of the novel material, FGM for marine risers are discussed. Material composition is chosen based on the functional requirements without compromising on the strength criteria, and FGM

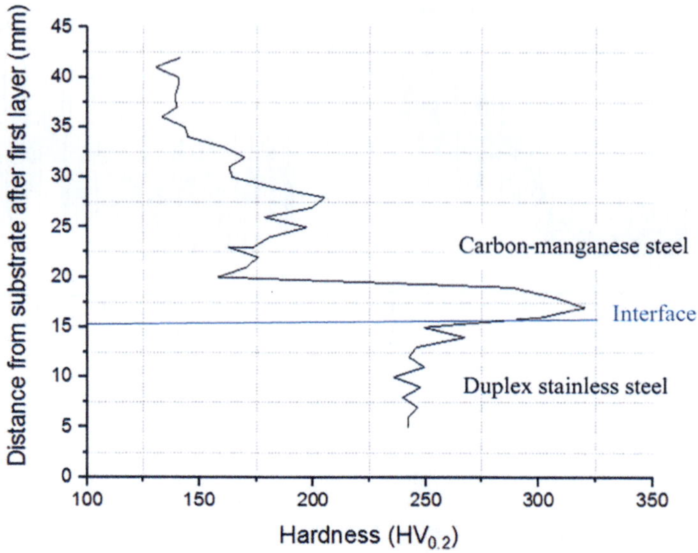

Figure 2.35 Variation in microhardness along the FGM build. *FGM*, Functionally graded material.

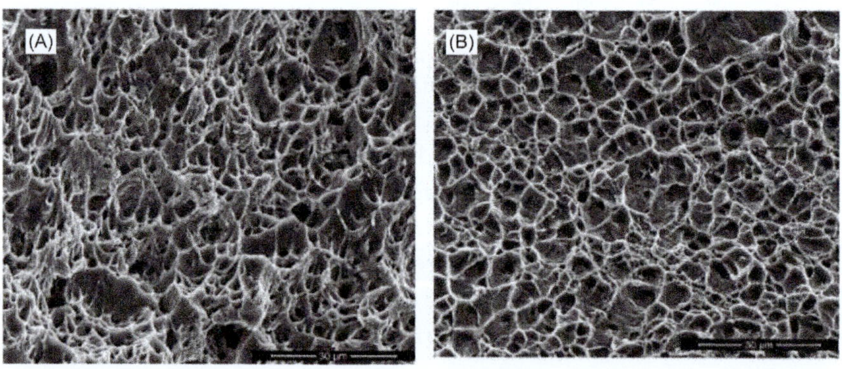

Figure 2.36 Fractography obtained at the (A) duplex stainless steel and (B) carbon–manganese steel.

samples are manufactured by WAAM. The CMT process is advantageous in terms of additive manufacturing due to its lowering of heat input and droplet detachment. The X-ray computed tomography results showed that the manufacturing process is structurally sound and well suited for part fabrication. From the comparison of mechanical properties, it is seen that the yield and ultimate strengths of FGM are higher by 11.16% and 24.75%, respectively, in comparison with that of the standard X52 steel

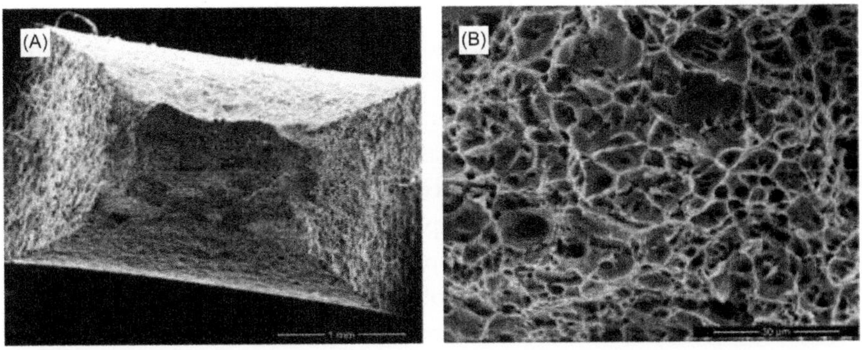

Figure 2.37 Fractography obtained at the failed portion of carbon–manganese steel (E8 subsize): (A) SEM image of the fractured portion of the tensile specimen and (B) fractography showing ductile failure.

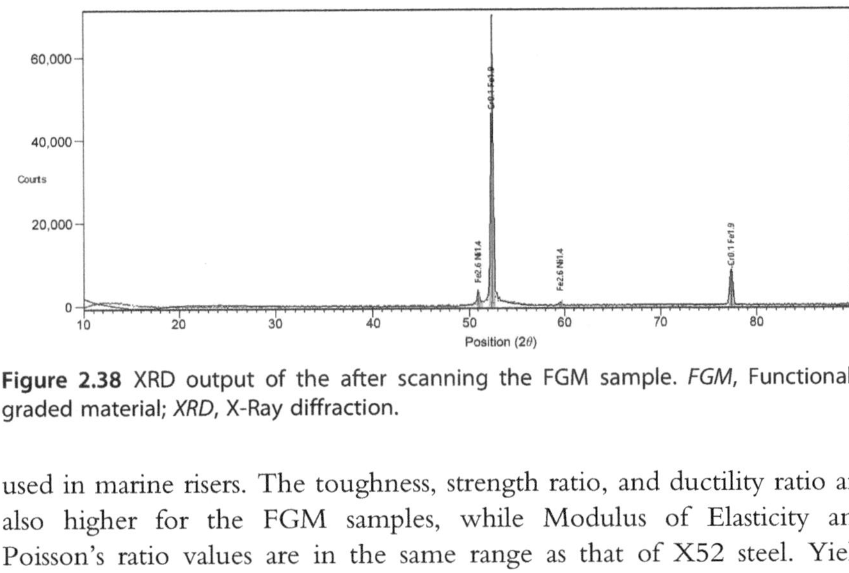

Figure 2.38 XRD output of the after scanning the FGM sample. *FGM*, Functionally graded material; *XRD*, X-Ray diffraction.

used in marine risers. The toughness, strength ratio, and ductility ratio are also higher for the FGM samples, while Modulus of Elasticity and Poisson's ratio values are in the same range as that of X52 steel. Yield strength of about 400 MPa is achieved both in the tensile and compressive direction during tensile tests, which makes it useful for axial and compressive loads acting on the marine riser. The microstructure analysis shows complete fusion between the two materials without any defect at the interface. The high strength zone is formed near the interface at the carbon–manganese steel side due to the martensite formation.

Apart from satisfying the strength criteria, marine riser with FGM under this combination shall also exhibit an excellent corrosion resistance due to the innate characteristic of a duplex stainless steel. It is validated by the EDX test, which shows that there is minimal diffusion of chromium across the interface toward the carbon–manganese steel side. Also, from the pitting corrosion test, there is a 12 times improvement in the corrosion rate with FGM as the material when compared with carbon–manganese steel (X52 Steel). The fractography analysis shows that the FGM material has undergone a ductile failure. There is a higher contribution to ductility from carbon–manganese steel. The present combination of FGM shall be useful in the future of marine riser applications.

From the stress–strain curves obtained during the tensile tests on FGM mould, it is observed that there is an increase of about 11.16% and 24.75% in the yield and ultimate strengths of FGM mould in comparison to that of the X52 marine steel. The ductility ratio, strength ratio, and toughness are also higher for the FGM material, while the Poisson's ratio and modulus of elasticity are in the same range as that of X52 marine riser steel. From the interface tensile tests, it is seen that the yield strength and percentage elongation are comparable to that of X52 material, while there is a marginal increase in the ultimate strength of FGM. Microstructure studies show that there is a complete fusion between the two-graded materials without defects at the interface. Martensite formation at the carbon–manganese steel face results in the formation of a high strength zone at the interface. From the XRD results, it is concluded that the resulting FGM combination is corrosion resistant, and the oxide layer is intact.

Marine riser with the present combination of FGM exhibits excellent corrosion resistance due to the innate characteristics of duplex stainless steel. It is further validated by the EDX test, which confirms that there is minimal diffusion of chromium across the interface toward the carbon–manganese steel. The fractography analysis showed that the FGM material had undergone ductile failure. There is a higher contribution to ductility from carbon–manganese steel. The present combination of a functionally graded combination shall be useful in marine riser applications in the future.

Exercise

Part A: Objective questions

1. ________________ (Functionally graded materials) are those materials whose structural properties are varied along their volume to perform an intended function.

2. Why are marine risers subjected to corrosion?

   Marine risers are subjected to severe corrosion due to the presence of chlorides, hydrogen sulfides, and carbon dioxide gases.
3. Composites are subjected to ___________ (delamination) under the combined action of both the mechanical and thermal loads.
4. How composites delaminate?

   Under the high-temperature and high-pressure (HT—HP) conditions, thermal stresses develop at the bonding interface causes delamination. Thermal stresses occur due to the difference in the coefficient of thermal expansion between the composition of materials.
5. ___________ (FGM) possess no distinct interface between the individual materials.
6. What are the factors that affect the strength of FGM?

   The strength of the FGM depends significantly on the manufacturing process. One should ensure a reliable interface between the constituent materials to meet the strength of the application desired.
7. Cold metal transfer is a modified form of __________ (gas metal-arc welding process), involving low current and voltage.
8. Name two metals that show high-corrosion resistance to hydrogen sulfide gases, chlorides, CO_2, and acidic environment (duplex stainless steel and titanium).
9. ___________ (stress corrosion cracking) is caused by chlorides in the marine environment
10. What are the various methods of manufacturing FGM?

    WAAM manufactures FGM. Other conventional methods of producing FGM, namely, centrifugal casting, spark plasma sintering, laser deposition, and powder metallurgy
11. Name a few additive manufacturing processes to manufacture FGM.

    A few additive manufacturing techniques for manufacturing FGM are, namely, selective laser melting, direct laser melting process, electron beam melting, and wire arc additive manufacturing, are also in practice.
12. What is the most significant advantage of WAAM?

    WAAM overcomes one of the severe limitations of the powder-bed process of manufacturing in terms of the deposition rate, equipment cost, and the cost of raw material.
13. ________________ (carbon—manganese steel) is one of the most commonly used materials for manufacturing marine risers.

14. What is the desired wall thickness of a marine riser?

$$t_{\min} = \frac{P_i D_{\text{nom}}}{2\,\eta_h \sigma_y} + t_{\text{cor}}$$

15. ________________ (Poisson's ratio) is estimated by measuring the longitudinal tensile strain and the transverse compressive strain.
16. ________________________ (energy dispersive X-ray analysis) is carried out to determine the elemental composition across the interface.
17. Draw a typical stress–strain curve of an FGM in the longitudinal direction.
18. Draw a typical stress–strain curve of FGM in the transverse direction.

Part B: Descriptive questions

1. Write a brief note on FGM.
2. Discuss WAAM process, its advantages, and limitations, if any.
3. Discuss the typical sectional properties, recommended by international codes for marine risers.
4. Write short notes on X-ray tomography, discussing its application to assess the characteristics of FGM.
5. Discuss in detail the tension test carried out on FGM.
6. Discuss in detail the microstructure analysis carried out on FGM.
7. Write short notes on energy dispersive X-ray analysis, explaining its use in assessing the characteristics of FGM.
8. Write short notes on microstructure analysis, explaining its use in determining the characteristics of FGM.
9. Write X-ray diffraction analysis, explaining its use in assessing the characteristics of FGM.
10. State a few advantages that FGM possess as a material for marine riser.

CHAPTER 3
Vortex-induced vibration

3.1 Introduction

When a liquid medium intercepts the bluff body, it will be influenced by vortex-induced vibration (VIV). It results in problems of twofold: (1) destructive overstress on structural members due to larger amplitudes and (2) strong potential to cause long-term effects such as fatigue failure (Sarpkaya, 1978, 2004). Both are of great interest to offshore design engineers. Marine risers are candidates to VIV, as they are slender. Their response is more complicated due to the reattachment of the vortices to the oscillating slender bodies. VIV is the outcome of such exciting, periodic forces, causing a response closer to the resonance period. Being undesirable, suppression of VIV shall improve the safe and comfortable working environment of offshore drilling by enhancing the design life of structural members.

3.2 Vortex-induced vibration

When a body is placed in a fluid medium, a change in pressure occurs on liquid particles, which alters the flow pattern, post the body interference. There will be an increase in pressure from the free-flowing condition to cause stagnation pressure. As the flow progresses and reaches a particular point, the boundary layer is separated from the bluff body, resulting in the formation of two shear layers on either side of the body. The location where the shear layer has contact with the body surface will be moving very slowly in comparison to that of the outer part of the shear layer, which maintains a free-stream velocity. As they approach the near-wake region, shear layer rolls to form vortices. They get separated from the flow regime and move to the wake region. During this process, vortices will impose additional forces on the encountering bluff body, causing excessive vibrations, termed VIV.

The boundary layer is a thin layer confined to the surface of the body. Viscous effects are very prominent inside the boundary layer. The flow will be frictionless outside the boundary layer. The presence of shear force will make the boundary layer flow rotational, whereas the flow outside remains irrotational; velocity will be of that of a free flow. Flow separation

Design of Marine Risers with Functionally Graded Materials
DOI: https://doi.org/10.1016/B978-0-12-823537-9.00007-4

from the body is well understood, explained with the help of pressure distribution around the body. Pressure gradient varies with the change in the flow velocity; pressure increases with the decrease in velocity. Hence, the flow velocity is vital to examine flow separation. Further, it also depends on the geometric shape of the encountering body, which changes the flow characteristics significantly.

Following are the pressure distribution for different cases:

1. When the flow is attached to the body surface:

$$\frac{dP}{dx} < 0 \tag{3.1}$$

2. When the flow is about to separate:

$$\frac{dP}{dx} = 0 \tag{3.2}$$

3. When the flow is separated:

$$\frac{dP}{dx} > 0 \tag{3.3}$$

As flow approaches the free surface, pressure varies, as shown in Fig. 3.1. Pressure will reduce as it flows through the surface of the body surface toward its right side, as seen in the figure. At a point P, it shall develop no pressure at the surface of the sphere. Due to the presence of an adverse pressure gradient, flow cannot stay attached to the surface of the body. It, therefore, gets separated and moves to the wake region as vortices. Momentum equation given by Navier–Stokes shall govern the phenomenon:

$$u\frac{\partial u}{\partial s} = -\frac{1}{\rho}\frac{dp}{ds} + \nu\frac{\partial^2 u}{\partial y^2} \tag{3.4}$$

where s and y are stream-wise and normal coordinates, respectively.

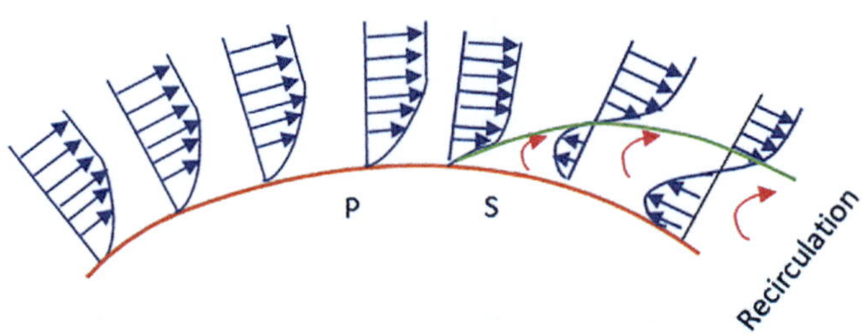

Figure 3.1 Pressure distribution and boundary layer separation.

Vortices are shed from bluff structures in subsonic flow. The flow characteristics are different according to different Reynolds number conditions. The vortex shedding frequency approaches the natural frequency of the structure with the change in flow velocity. Thus lock-in occurs, and structure starts vibrating with higher amplitude. Different parameters, such as Strouhal number, reduced velocity, amplitude ratio, etc., are essential in the study of VIV (Blevins, 1994). Vortex shedding is a common phenomenon for all flow regimes of $Re > 40$. The boundary layer separation from the structure is mainly due to the variation in the pressure gradient. Vorticity formed in the boundary layer is transferred to the shear layers, which causes the rolling up of layers and form separation. Mutlu and Jorgen (2003) explained how normalized vortex shedding frequency (which is represented by Strouhal number) is affected by various flow velocities. The Strouhal number has a constant value throughout the entire subcritical range of flow regime. If the flow velocity T_s further increased to critical and supercritical flow, a sudden rise in this value can be seen.

3.3 Vortex-induced motion

When Reynolds numbers (Re) exceed 40, the mechanism of vortex shedding changes, as shown in Fig. 3.2. As seen in the figure, two unstable pairs of vortices are formed around the cylinder. While they are subsequently subjected to small disturbances, both vortices will not shed at the same time. One of

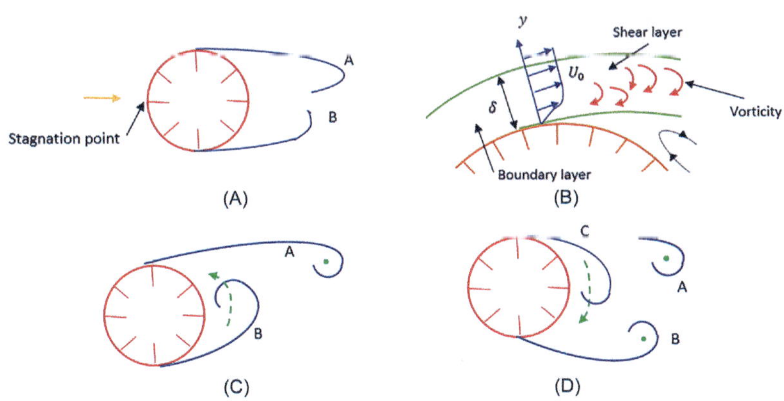

Figure 3.2 Mechanism of vortex shedding: (A) shear layer on both sides, (B) flow separation from the surface, (C) vortex B, pulled into wake region, and (D) vortex B sheds to draw vortex C into wake.

them is expected to grow more significantly than the other. Larger vortex, as marked A in the figure, will be stronger and, therefore, possess a tendency to pull vortex B across the wake region; the direction of both the vortices is opposite. While vortex B approaches the wake region, vortex A will cut off and results in shedding. After shedding a new vortex (vortex C) will be developed on the same side. More significant growth of vortex B will result in pulling vortex C, causing shedding of vortex B, and this process will be repeated. It results in shedding of vortices alternatively from the cylinder, as shown in the figure.

Many factors influence the vortex-induced motion (VIM), namely, amplitude, range of current velocities, the cross-sectional shape of the member, ratio of the body mass to the displaced fluid mass, and damping. Considering the importance of offshore structures, one has to recognize such uncertainties and should address them adequately in design (Chandrasekaran, 2015a). Marine riser, under free-floating conditions, indicates natural surge and sway periods as 2−3 min. As this is highly flexible, causing a slow vibration, this motion is termed VIM, rather than vibration, although the physics are the same. These motions have at least two possibly critical effects, namely. In essence, repeated cycling can induce fatigue damage to risers and can increase the apparent hull drag coefficient, resulting in unacceptable mooring loads under high currents. VIV suppression devices such as helical strakes are deployed to mitigate VIM. While helical strakes are known to work on smokestacks and risers, they are found to negate the effectiveness of strakes due to their effect on the fluid boundary layer around the member.

Vortex shedding of a riser, under steady flow, depends on Reynolds number. It is a function of stream velocity (*U*), the diameter of the riser (*D*), and the kinematic viscosity of the fluid (v), and is given by the following relationship:

$$Re = \frac{UD}{v} \tag{3.5}$$

At lesser values of Reynolds number (e.g., $Re < 5$), water moves along the surface of the cylinder. Upon increased value between 5 and 40 a symmetric pair of vortices is shed as the flow separates from the back of the riser. Further increase in the Reynolds number increases the stream-wise length of the riser. For example, at 45, the length of the member will be about thrice of its diameter. Wake becomes unstable on a further increase of the Reynolds number, leading to the formation of vortices.

Such vortices shed alternatively on both sides of the riser. When the Reynolds number lies in the range (40–200), a laminar vortex street will be observed. However, shedding does not vary in along the span of the riser, as it remains two-dimensional. At a higher range of the Reynolds number, say in the range of (200–300), a laminar flow will transit into a turbulent flow in the wake region, causing the turbulence formation near the vicinity of the cylinder (Bloor, 1964). At this stage the two-dimensional properties observed earlier will change into three-dimensional (3D) (Gerrard, 1966; Williamson, 1988), which means that there will be vortex shedding along the span of the riser, causing the wake region turbulent.

The range of the Reynolds number ($300-30 \times 10^5$) is termed a subcritical flow regime. Under these circumstances, turbulence shall start to appear at the boundary layer itself. It appears initially at a point where the flow gets separated from the surface of the riser and then shifts to the upstream side toward the stagnation point. The boundary layer becomes turbulent at point of separation within a narrow band of Reynolds number ($3-3.5 \times 10^5$). While the turbulence will be noticed only on one side of the cylinder, the other side shall exhibit a laminar flow. This regime of flow is termed the critical or lower transition regime. The range of the Reynolds number ($3.5 \times 10^5-1.5 \times 10^6$) is termed the supercritical regime in which the flow experiences separation of the boundary layer on both sides. However, a complete transition into the turbulent region does not happen at this stage. A transition from the laminar to turbulent flow shall occur between the stagnation and separation points. When the Reynolds number exceeds (1.5×10^6), one side of the cylinder (riser) will experience a turbulent flow while the other side shall continue to experience partly laminar. This type of regime is termed an upper transition flow regime and shall prevail over a range ($1.5 \times 10^6-4.5 \times 10^6$). Upon further increase of the Reynolds number beyond (4.5×10^6), the entire boundary layer will become turbulent, and this regime is termed a transcritical flow regime.

Experimental studies carried out on response analysis of elastic-mounted cylinders under VIV showed that there exists a cross-flow at lower Reynolds number (90–150) (Anagnostopoulos and Bearman, 1992). In this regime a fully laminar vortex street is observed. The velocity of the fluid in the cylinder wake region showed maximum amplitude near the vicinity of a lower limit of the lock-in region. The cylinder oscillates at a frequency higher than its natural frequency in the lock-in region. Khalak and Willaimson (1991) identified that the mass ratio is one of the parameters that influence the regime of synchronization. Govardhan and Williamson

(2000) examined the modes of a freely vibrating cylinder and measured the vorticity, force, and displacement. They observed a jump in the vortex phase when the frequency becomes closer to that of the natural frequency. The jump is associated with two different wake modes, which are referred to as 2S and 2P. At a higher mass ratio (=100), vibration frequency stays closer to that of the natural frequency but reached maximum amplitude.

Chandrasekaran and Madhavi (2015b) investigated on the feasible structural modifications that resulted in a significant reduction of active forces on a cylindrical member. They studied the effect of an outer perforated cylinder that covered the inner damaged member. It was one of the successful attempts to reduce the force induced by waves. Chandrasekaran et al. (2015) also conducted experiments on the tension leg platform (TLP) model with a tuned mass damper. It was effective in the reduction of the motion of the entire platform. Hence, recent studies prompted toward response (and force) reduction on the cylindrical member in the fluid medium, when they experience VIV.

3.4 Lock-in phenomenon

When the natural period of the member synchronizes with the shedding period, the member will lock-in to vortex shedding frequency, termed lock-in phenomenon. Frequency of the vortex shedding, which is otherwise called as Strouhal frequency is dependent on the Strouhal number (*S*) and is given as follows:

$$f_s = \frac{S\,V_c}{D} \tag{3.6}$$

where V_c is the current velocity and D is the diameter of the cylinder member. An inverse of the Strouhal shedding frequency is termed Strouhal shedding period. The natural period of the riser depends on its submerged mass and stiffness of the mooring lines, supporting the riser, and is given by the following expression:

$$f_n = \frac{1}{2\pi}\sqrt{\frac{K}{(M + M_a)}} \tag{3.7}$$

$$T_n = \frac{1}{f_n} \tag{3.8}$$

where K is global lateral stiffness at the mean offset position, M is mass of the body, and M_a is the added mass. Added mass is computed by testing a scaled model of the riser in an anchored position. In the case of risers or pipelines, Strouhal number is a function of the Reynolds number as the surface roughness, and free-stream turbulence does not influence the Strouhal number. Fig. 3.3 shows the range of both the numbers; it can be seen from the figure that over a large range of Reynolds numbers, the Strouhal number is about 0.2. The following empirical formula can be used when the Reynolds number is between $250 < Re < 2 \times 10^5$:

$$S = 0.198\left(1 - \frac{19.7}{Re_D}\right) \tag{3.9}$$

Lock-in is related to the reduced velocity and is given by the following expression:

$$V_m = \frac{V_c T_n}{D} \tag{3.10}$$

where V_m is obtained from the velocity at mean offset position under the combined action of the wave, wind, and current, which is determined by an iterative process. The amplitudes of both VIV and the drag force influence the offset position significantly. Therefore lock-in is considered a local synchronization of both the frequencies of the member and the vortex shedding. In such an analysis the amplitude (A) in the transverse vibration is considered a nondimensional number. It is computed as the ratio of the amplitude in sway direction to its diameter (A/D). For unit value of this ratio the largest possible single amplitude occurs in the transverse direction on a bluff body. However, under certain circumstances, this ratio is bye-passed as the response amplitude for different members is not

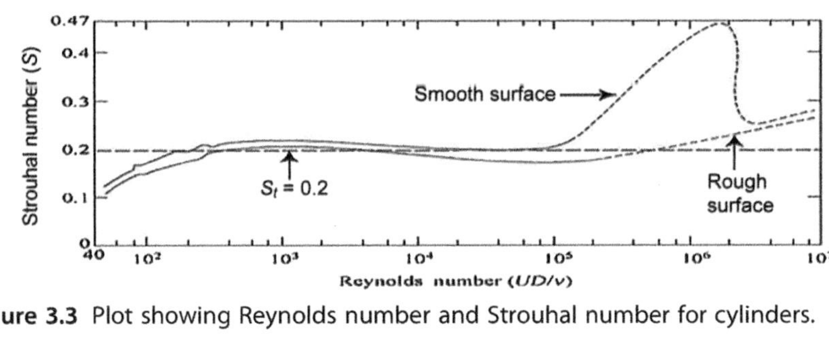

Figure 3.3 Plot showing Reynolds number and Strouhal number for cylinders.

identical. It depends on other factors, namely, geometry and shape of the riser, profile of current, presence of vortex-mitigation devices, other appurtenances fitted to the riser, mass ratio, and damping (Feng, 1968). Besides, an asymmetric mooring arrangement also affects the trajectory of the VIV. In such cases the major axis about which VIV occurs does not remain normal to the direction of the current.

The amplitude of VIV in the transverse direction varies with the variation of the reduced velocity. The response of the cylindrical member shall be insignificant if the magnitude of the reduced velocity is lesser than that of the threshold value (which is usually 4). For this particular range of reduced velocity, the response of the cylinder under VIV attained its maximum amplitude and termed lock-in range. Lock-in typically occurs in the range of the reduced velocity (4–7) but sometimes can be as high as 10. Beyond this range of the reduced velocity, amplitude under VIV reduces, which is termed lock-out region. But the lock-out region is highly inconsistent as the range of this value can be very large in some cases. Based on the geometry of the member, lock-in may not even occur. There also exists a possibility to result in a reduced velocity of larger magnitude, causing galloping (Parkinson, 1989). Fig. 3.4 shows the plot variation of reduced velocity and the amplitude ratio of VIV; both lock-in and lock-out regions can be seen in the figure. As the risers are rigidly connected only to the top of the platform, VIV depends on the direction of the current; the bottom of the riser remains free-floating. For

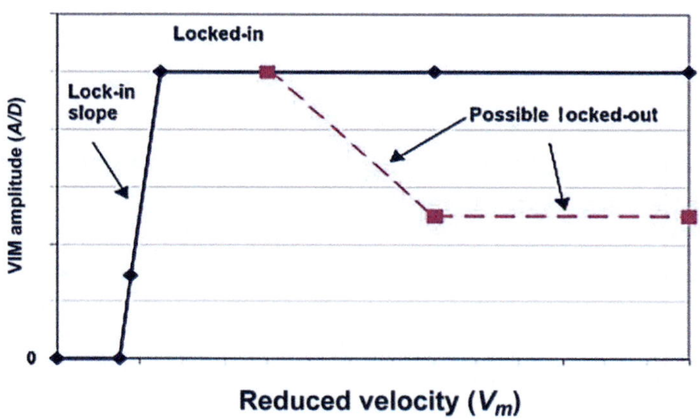

Figure 3.4 Amplitude of VIV on different reduced velocity. *VIV*, Vortex-induced vibration.

the same amplitude of the reduced velocity, response amplitude under VIV changes with a change in the direction of the current. It is in addition to the effect of the geometric shape of the risers.

3.5 Inline vortex-induced vibration

Drag forces can generate inline effects on the member encountering wave loads. Inline drag forces occur along the direction of the current and influence the amplitude of the VIV in the transverse direction. Inline VIV is also dependent on other parameters, namely, the existing profile, geometric shape of the cross section of the member, and material (Chandrasekaran and Jain, 2016). Other parameters, such as mass ratio, damping, and appurtenances fixed to the member, also affect the vibration amplitude (Chandrasekaran, 2017). But the magnitude of inline vibration is lesser than that of the amplitude in the transverse direction. It is observed that inline vibration of an offshore floating structure like classical spar classic spar with symmetric mooring is about 10%–15% of that of the vibration in the transverse direction. It should be noted that the amplitude of inline response can exceed that of transverse response when the natural period of motion synchronizes with that of half of the natural period of the transverse motion (Chandrasekaran and Merin, 2016). Different factors that affect inline VIV are current velocity and direction, duration, and profile of current; most common are tidal and circulation current.

In the real sea, marine structures are subjected to current and waves of higher magnitude. A combination of low-frequency wave and wind, acting normal to the direction of the current, increases the transverse amplitude of VIV. Hence, the structural design should include the combined effect of waves and current as marine risers are sensitive to large amplitude waves (Chandrasekaran, 2017). A significant number of special issues should be addressed to assess the impact of VIV on floating offshore platforms and flexible risers. Model testing is one among them to estimate the design criteria in such cases. As field data for different situations are quite challenging to obtain, model tests are used as useful substitutes to simulate field measurements. For getting meaningful results, various parameters such as current velocity, profile, the shape of the hull, and geometry should be appropriately accounted in the model. The period of extreme current and the corresponding VIV can be reasonably more extended than those occurring under extreme storm duration. In such cases an appropriate statistical tool can be used to predict acute VIV responses.

A few numerical simulations that are commonly used to estimate VIV are, namely, SHEAR7, VIVA, and VIVANA. Most of them are developed based on either empirical or semiempirical relationships for fluid–structure interaction. Most of these simulation packages use modal analysis tools to obtain approximate solutions. Alternatively, a program such as Flexcom-3D ensures a proper coupling between the structural response and fluid loading. A few of the computation fluid dynamics tools are also available to solve the classic Navier–Stokes equation, but their solutions demand a longer simulation time. In addition to this, high-end calibration is also required to estimate the drag coefficient under a higher Reynolds number (Chia and Hamn, 2016). In some instances, due to the geometric configuration of the risers, analysis becomes important to simulate in 3D space. Hence, it is better to use a finite element Eigen frequency solution program to generate shape, curvature, and mode frequency for the system. Further, this modal data can be used as a vital input for assessing the VIV response.

Before designing a model test needs to be conducted to determine the drag coefficient (C_d). In the case of a zero-amplitude vibration ($A/D = 0$), one can assume an initial base drag (C_{d0}). An amplification factor (k) is applied to include VIV effects. Drag augmentation is expressed as a function of parameters (A/D, V_{rn}) and is given as below:

$$C_d = C_{d0}\left[1 + k\left(\frac{A}{D}\right)\right] \tag{3.11}$$

where C_{d0} is the initial drag coefficient for the condition (VIV = zero), and k is an amplification factor calculated through model tests. Drag force of cylinder is given by:

$$F_d = C_d\frac{1}{2}\rho V_c^2 DL \tag{3.12}$$

where C_d is the drag coefficient and ρ is the density of the fluid. For a sheared profile of current, drag force is obtained by integrating over the entire length of the hull and is given as follows:

$$F_d = C_d\frac{1}{2}\rho\int_{-L}^{0} V_c^2(z)D\,dz \tag{3.13}$$

For offshore structures with members of different diameters, D, in the above expression, is considered as the diameter of the hard tank. Drag force of various components of truss members is added to obtain the total drag force of a typical truss spar.

3.6 Suppression of vortex-induced vibration

When deepwater risers are analyzed for VIV, it is mandatory to include the effects of long-term and extreme short-term events; it is mostly generated in the presence of current. However, waves also contribute to an increase in the amplitude of VIV. In addition to these, wave-induced heave motions also play a significant role in the VIV generation. Long-term effects of VIV include fatigue damage. It is important to note that extreme events are to be included in the analysis as they can cause sudden damage to the risers. Apart from the surface currents, it is important to map the full profile of a water column. In most cases of field developments, some conservative assumptions are applied wherever field data is insufficient. In the field where high currents are experienced (see, e.g., current velocity >2 kn), risers are more prone to VIV. One example of such a region is the Gulf of Mexico (GoM). In GoM, there exist cold-core eddies and loop currents as a result of high-current profiles. Loop currents are generally present on the near-surface, whereas maximum cold-core eddies are present at a depth around 400 m below the mean sea level. When maximum velocity prevails at a higher depth, longer sections of the riser are exposed to current and thereby experience fatigue damage substantially.

There are primarily two approaches to address the issues related to VIV in risers. Either the design or geometry is modified to eliminate VIV, or appropriate VIV suppression devices are to be used as external attachments to the riser. Among both the alternatives, the former is a better idea as it reduces the need for suppression devices. However, design changes are not practical in all types of risers. For example, in the case of a steel catenary riser, the VIV suppression system includes strikes with an increased hang-off angle. But this will alter the tension distribution in the riser and further modifies the natural frequencies and modal responses. In the case of types of risers where design modifications are not feasible, VIV suppression devices are commonly deployed to disrupt the formation of vortices. A few of the VIV suppression devices include strakes, fairings, shrouds, and dimples. Among them, the most commonly used are helical strakes and fairings. The helical strakes consist of fins, which are wound helically around the surface of the riser. It is similar to the technique adopted in the stacks of tall chimneys. Helical strakes can interrupt the shedding of vortices and hence weaken the vortex formation. However, it increases the drag significantly, and this should be accounted for in the design. It adds to an additional problem of attracting marine growth.

Fairings are aerofoil-shaped elements that can control the flow pattern around a body. It can streamline the flow, which results in the reduction of the strength of vortices formation (Evan, 2007). Fairings are found to be effective only when they are oriented toward the stream direction. Therefore they should be designed to remain free to rotate and cause "weathervane" around the member. Though there are some potential advantages of fairings over the strakes, the latter is the most common suppression device employed on the marine risers. It is because fairings demand moving components, which are capable of getting oriented to the current prevailing direction. They are considered impractical for the typical design life requirements of the risers. But, for example, fairings deployed on the Mars TLP are found to be successful even after 10 years in service.

Among various methods that are proposed to suppress VIV, strakes and shrouds are found to be effective regardless of the orientation of structure to waves and current but showed a substantial increase in drag, resulting in high drift. It seems to be responsible for an increase in the stiffness of the anchoring system. Despite the higher cost, helical strakes are recommended to strengthen the cylindrical member against larger bending moments that arise from increased drag. The strake design is verified and optimized using experimental investigations since the VIV behavior of cylinders cannot be predicted precisely using numerical tools. Zdravkovich (1981) conducted studies on different aerodynamic and hydrodynamic methods to mitigate VIV. They address shrouds, Strakes, Fins as different means of suppression systems. Owen and Masa (2003) conducted experimental investigations with passive suppression devices. They fitted the cylinder with strakes and bumps. Under the deployment of a single splitter, the steady transverse force was observed. When strakes or bumps are fixed, no regular shedding occurred on the wake side of the cylinder. But, when the mass-damping parameters are reduced, traces of regular vortex shedding are observed at a particular range of reduced velocity. Nevertheless, the amplitude of vibration of the cylinder is suppressed effectively by deploying the strakes and bumps. Evan (2007) explored the means of suppressing VIV of an isolated cylinder having low mass ratio and mass damping. When a splitter plate of length equal to the diameter of the cylinder was fitted with an allowance to rotate free, effective suppression was observed at an angle of 20 degrees from the stagnation point. A maximum suppression was noticed by deploying two plates at an included angle of 40 degrees from stagnation point; this arrangement reported the absence of any steady force.

Various methods to minimize the effects of vortices are also employed by Ricardo (2008); Delta wings are utilized to generate vortices. They are fixed parallel to the free stream to ensure the formation of vortices, parallel to the streamflow. Results showed a major suppression at a lower amplitude. Delta wings of smaller dimensions showed considerable suppression along with the drag reduction. A modified configuration of the delta wing, termed Rabbit Ear, which is half of the size of the delta wing, showed an effective suppression. Different studies on suppression systems using airfoils at different angles also proved to be effective in VIV suppression (Williamson and Govardhan, 2004). Depending on the relative position of the airfoils, flow-induced motion is altered. They are found to be effective in suppressing the vibration of multiple cylinders. Another effective method of suppressing the vortex-induced forces and motion by controlling the surface roughness. When the added roughness is designed and executed properly, boundary layer thickness and separation, turbulence level, wake region, lift forces, and drag forces are affected, more or less, in a predicted manner (Park et al., 2012, 2013).

3.7 Flow characteristics

The behavior of fluid flow that encounters the bluff body is an essential area of investigation. Studies in VIV are initially conducted where the circulating medium is taken as air. Flow visualization done by laser sheet—based smoke showed how the separation point changed along the surface of a cylinder (Lesage and Garthshore, 1987). Govardhan and Williamson (2000) studied the wake properties through the flow visualization behind a pair of cylinders that are placed side-by-side. It is seen that in the case of phase synchronization, which took place through the size of the gap between the pairs of cylinders, two vortex streets, referred to as binary vortex, are formed. They also confirmed that the formation of a couple of antiphase streets and in-phase shedding could result in a single wake of large amplitude. It is interesting to note that the gap between the cylinders significantly influenced the characteristics of wake properties. Asymmetric waves are formed if the spacing between the pair of cylinders is lesser than that of the critical value. It causes a hot-wire frequency in terms of harmonic modes. It is also important to note that pair of vortices, which are shedding alternatives, produces a force that makes the cylinder vibrate in the transverse direction. Different trends existing between Strouhal number, Reynolds number, and amplitude ratio are studied experimentally on flexible cylinders by various researchers recently (Resvanis et al., 2012). They showed that flow separation on cylinders is primarily affected by their length. Also, both their diameter and boundary layer thickness.

3.8 Behavior of flexible members

Slender structural members such as mooring lines and risers are affected by the VIV. In such structures, steady current can cause vibrations of larger amplitude, which will intuit fatigue damage. Such repeated stress cycles lead to an early failure of risers (Chakrabarti, 2002). Griffin and Ramberg (1982) discussed the effects of vortex shedding on drilling risers and quantified adverse effects caused by increased drag, in particular. Recent studies also showed the VIM on long flexible cylinders (Griffin et al., 1973; Brika and Laneville, 1993). They noted a hysteresis loop motion of the cylinder under VIM in which each of the vortex shedding modes contributed to the loop. There was a jump from one branch to another, causing an abrupt change in the frequency of the vortex shedding. Under constant velocity of motion of the cylinder, both impulsive and progressive regimes are observed. In an impulsive regime, bifurcation coordinates defined the boundary of modes of vortex shedding. Similar effects were also reported on long, flexible cylinders; most importantly, change in the Reynolds number of the flow field influenced the frequency of vortex shedding significantly (Hang et al., 2003; Susan et al., 2008).

3.9 Vortex-induced vibration suppression: Experimental investigations

In this section, experimental studies carried out to suppress VIV are discussed. As discussed earlier, model tests have a serious limitation in terms of scale ratio. Some of the parameters should be addressed while modeling VIV. They are, namely, geometric similitude, hydrodynamic scaling, dynamic similitude, and degrees-of-freedom of the scaled model. Geometric similitude implies the similarity in shape. A prototype cylinder of 22 m in diameter and 215 m in length is modeled as a cylinder of 0.102 m in diameter and 1 m in length. Reynolds scaling and Froude scaling are two significant scaling methods for hydrodynamic model testing. Both these numbers are expressed in the following equations:

$$Re = \frac{V_c D}{\nu} \tag{3.14}$$

$$F_n = \frac{V_c}{\sqrt{gD}} \tag{3.15}$$

where Re is Reynolds number, F_n is the Froude number, V_c is the current velocity, D diameter of the member, and ν is the kinematic viscosity of the fluid. Simultaneous scaling using both the Reynolds and Froude's may not be practically possible for floating bodies under the effect of current. Either a substantial increase in gravity or a decrease in viscosity is required for the model dimension. It is significantly different in comparison to that of the prototype. For a typical floating platform connected with risers, normal sea states show the presence of current velocity in the range (1–2.5 m/s); hence, Froude scaling is followed in the current investigation.

Dynamic similitude ensures the similarity of rigid-body dynamics, and it is related to the reduced velocity and mass ratio. Even though modeling all of the modes of a rigid body may not always be essential, lock-in for different degrees-of-freedom does not need to occur at a particular velocity (Finn et al., 2003). For example, a floating cylinder may lock-in to sway at some particular velocity, which may be different from that of the lock-in velocity in the roll. In the current investigation a single degree-of-freedom system model is chosen. The mass ratio can induce a significant effect on the range of lock-in amplitudes and velocities. Therefore, in the current investigation, a unit mass ratio is maintained for the freely floating body. Experimental investigations under supercritical Reynolds number for a single degree-of-freedom model allow only VIV only in the transverse direction.

In the previous sections, it is well established that VIV response amplitude (A/D) is associated with the reduced velocity (V_m). Hence, the response amplitude of the cylinder is measured at different reduced velocities. It is important to note that the chosen range of reduced velocities for the experimental studies should include the reduced velocity, corresponding to the inverse of the Strouhal number (Bearman, 1984). It is because the inverse of the Strouhal number is closer to the peak amplitude of the reduced velocity. Further, it is also interesting to note that within the capture range of reduced velocities, the location of the maximum amplitude depends on the shape of the bluff body (Gabbai and Benaroya, 2005). In the current study the Strouhal number is taken as 0.2, and, therefore, the maximum amplitude expected shall occur around the range of the reduced velocity, which is equal to 5. Following expressions are useful to estimate the velocity of flow:

$$V_c = \frac{V_m D}{T_n} = \frac{5 \times 0.1}{0.63} = 0.73\ \text{m/s} \tag{3.16}$$

Hence, reduced velocity is considered as one of the main design parameters in the current investigation. An equivalent mooring line is modeled to represent the group of mooring lines, whose stiffness provides constraint against the transverse motion. At resonance condition, vortex shedding frequency will be equal to the natural frequency of the structure. Hence, the following expression holds good:

$$\frac{S\,V_c}{D} = \frac{1}{2\pi}\sqrt{\frac{K}{M}} \tag{3.17}$$

$$\frac{0.2 \times 0.73}{0.1} = \frac{1}{2\pi}\sqrt{\frac{K}{19}} \tag{3.18}$$

where V_c in m/s, D in m and M in kg, and K is computed as 1.6 N/mm.

The scaled model of a cylinder is vertically mounted, as shown in Fig. 3.5, and idealized as a single degree-of-freedom system model. Under the unidirectional waves the rest of the degrees-of-freedom are constrained except the sway motion. Mooring lines are represented a single spring attached to the top, as shown in the figure. Experimental studies are carried out in the subcritical range of the Reynolds number, with care to minimize the Froude number. Alternatively, the cylinder can be horizontally mounted, as shown in Fig. 3.6. It is important to ensure that the model experiences a uniform current along its entire length. A better idea would be to use a double body configuration, which has a splitter plate at

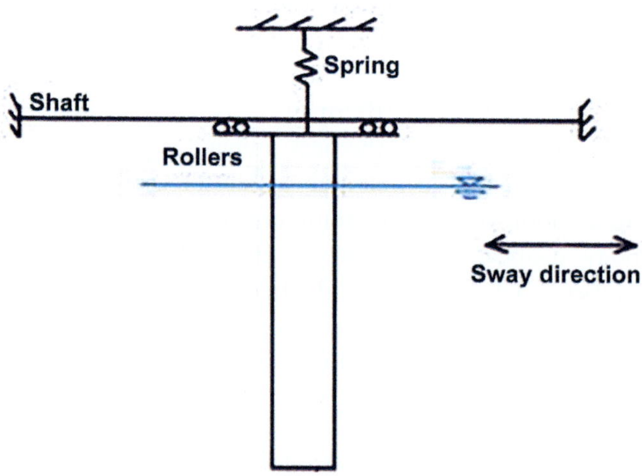

Figure 3.5 Experimental setup: vertically mounted model.

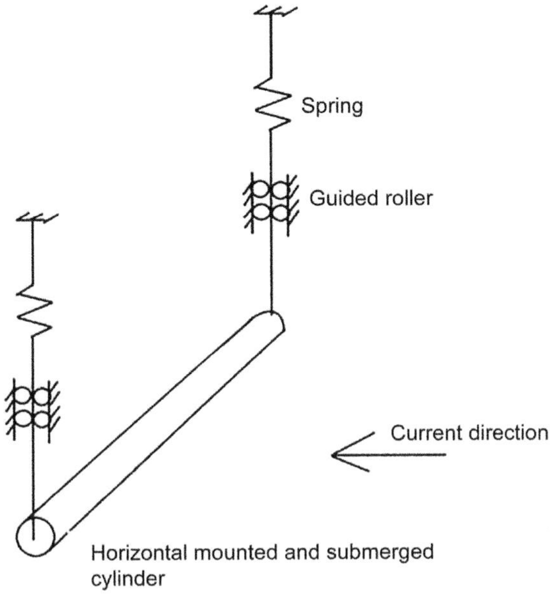

Figure 3.6 Alternate setup: horizontally mounted model.

its center as this can prevent the transfer of flow across the plate; it also helps as a check for data consistency.

The cylinder vibrates under the alternating forces developed due to vortex shedding. In the synchronization range, these force components are used to predict the dynamic response of the cylinder (Sarpkaya, 1977). Preliminary experiments are carried out to estimate the order of magnitude of these forces under fluid–structure interaction. As discussed in the previous sections, the inline force increases with the increase in amplitude ratio (A/D); the transverse amplitude is referred to as A. Other parameters that influence the response of the cylinder are viscous damping, the ratio of shedding frequency to natural frequency, and mass ratio (Chandrasekaran, 2015b); mass ratio is the ratio of the displaced mass of the fluid to that of the mass of the cylinder. At resonant conditions the peak amplitude of the oscillation depends only on a mass-damping parameter. It is also important to note that the structural damping decreases within the bandwidth of the lock-in frequency. Also, the maximum resonant response amplitude is indirectly proportional to the reduced damping.

Stripping wires are used in the current investigation to suppress the vortex shedding effects on the suspended cylinder. Effective suppression depends on the wire diameter and the flow pattern past the cylinder (Igarshi, 1986). Recent studies showed that three discrete flow patterns, namely, intuiting a laminar flow past the wire, forming a turbulent boundary layer immediately past the cylinder, and flow separation at the wire, can be considered. Out of the above three cases the second one resembles the naturally occurring critical flow at higher Reynolds number. In the present study the effect of stripping wires and their placement angles corresponding to the stagnation point are explored. Studies showed that stripping wires placed at the rear side of the cylinder would not have influenced the suppression of VIV (Hover et al., 1998). Hence, stripping wires are placed facing the incoming flow.

At higher Reynolds numbers ($Re > 10^5$), cross sections that are having fixed separation points, similar to those of the rectangular shapes, have a lesser influence on the vortex formation. But, in the case of circular shapes, a significant variation is noted with an increase in the Reynolds number. For $Re > 200$, vortex shedding occurs in the cells along the length of a cylinder. As the vortex shedding is not uniform along the length, forces acting on the range of a single cell shall be higher than the resultant force imposed over the entire length of the cylinder. To account for this uncertainty, correlation length, which is the average length of the cells are considered in the analysis (King, 1977). Table 3.1 summarizes the correlation length for a different range of Reynolds numbers (Gerlach and Dodge, 1970; E-Baroudi, 1960; Humphreys, 1960).

As seen from Table 3.1, for a specific range of Reynolds number ($40 < Re < 150$), the correlation length is found to be about (15–20) times of that of the diameter of the cylinder. Since the vortex regime is two-dimensional in this range, the correlation length is infinite, theoretically. Such a system with an infinite value cannot be obtained practically due to the boundary conditions. However, finite correlation length may

Table 3.1 Correlation lengths and Reynolds numbers of smooth cylinders.

Reynolds number	Correlation length
$40 < Re < 150$	$(15-20)D$
$150 < Re < 10^5$	$(2-3)D$
$1.1 \times 10^4 < Re < 4.5 \times 10^4$	$(3-6)D$
$\geq 10^5$	$0.5D$
2×10^5	$1.56D$

occur as a result of oblique shedding due to small variations from the pure two-dimensional shedding. For example, in the case of the flow with Reynolds number (=150), the laminar vortex shedding regime disappears. It shall result in a quick fall in the correlation length to about 2–3 times that of the diameter. It is interesting to note that the increase in oscillation amplitude increases with the increase in the correlation length.

3.10 Experimental setup

Experimental investigations are carried out in the towing tank of size $82 \times 3.2 \times 2.8\ m^3$ water depth (see Fig. 3.7). The velocities between 0.02 and 5 m/s are achievable in the setup. However, the presence of the test specimen and other minor operational issues impose constraints to run the tank at its maximum speed of 5 m/s. In the present study, vertically mounted cylinder created enough drag to limit the maximum achievable speed up to 1.2 m/s; hence, flow with a Reynolds number of 1.5×10^5 could only be tested, enabling the range of Reynolds number for testing as $(2.5 \times 10^4 - 1.5 \times 10^5)$. As explained earlier, within the chosen interval of *Re*, Strouhal number assumes a constant value of 0.2. Tests are carried out at 2.6 m water depth under uniform free-stream velocities.

In the entire study, water is kept stationary. Only the test specimen is moved at a different speed. Thus the relative velocity between liquid and the cylinder is maintained, and not physical fluid flow. Fig. 3.8 shows the test assembly. The test specimen is attached to the movable platform in the direction normal to its surface. This arrangement keeps the sample oriented

Figure 3.7 Towing carriage, used for the investigation.

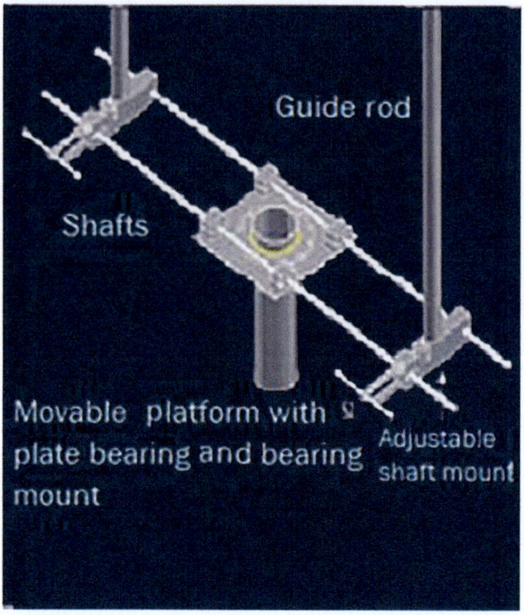

Figure 3.8 Test assembly.

perpendicular to the water plane and surface pierce. It ensures motion only along the direction perpendicular to that of the carriage motion and idealizes a typical single degree-of-freedom system model.

The test setup consists of a square plate of size $20 \times 20\ cm^2$, which is used as a platform to fix the cylinder vertically. To allow displacement in the cross-flow direction with a minimum possible friction, four linear motion bearings (Nippon LM20UU), which are sliding over two 20 mm diameter stainless steel shafts, are used. The bearings are clamped to a plate to create a movable platform. There are bearings mounted on each corner of the plate with horizontal shafts, as shown in the figure. The shaft consists of two horizontal rods placed parallel to each other, which enables the movement of the cylinder as it is fixed to the plate. Shafts are attached to the guide rods, which are mounted vertically to prevent the inline motion of the arrangement. The shafts are held by two adjustable mounts, which are fixed to a pair of guide rods. Suitable diameter and thickness of the guide rods are chosen to restrict the inline motion of the specimen. The guide rods are clamped to the carriage, as shown in Fig. 3.9.

Springs, connected to the top of the cylinder by a hinged joint, represents the mooring system. The total stiffness of 1.6 N/mm against the

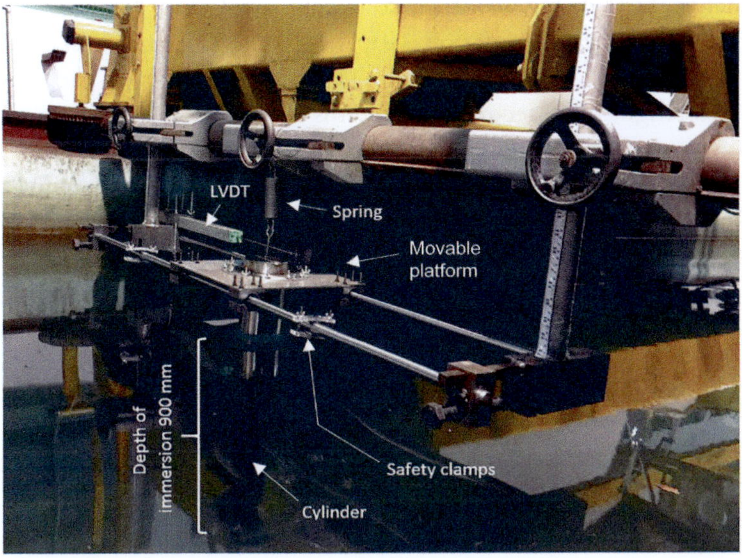

Figure 3.9 Experimental setup assembled to carriage.

transverse motion of the specimen is added to the network through these springs. The natural frequency of the models could be increased by increasing the initial tension in the spring. For the present study the natural frequency of the system is kept constant. Springs are, in turn, connected to the carriage facility by clamps. Linear variable differential transformer probes are connected to the plate where the cylinder is mounted to measure the motion of the cylinder; these probes can measure in the range ± 20 cm and are fixed using clamps. These clamps also ensure that the probes are not disconnected even under the extreme motion of the cylinder. A data acquisition system is connected to 12 V DC supplies to acquire the data.

The amplitude of VIM depends on the natural frequency and the damping ratio of the singe degree of freedom system. The system would lock-in when its natural frequency matches with that of the vortex shedding frequency. Free-decay tests in air and still water are performed for evaluating both the natural frequency and damping ratio. It is assumed that the viscous damping of fluid dominates over the frictional damping caused by the bearings. During the free-decay test, it is ensured that no twisting is developed in the model or the test plate. The cylinder specimen is fabricated using steel (Grade 304 steel), and the immersed ends of the cylinder are sealed by welding. Three samples are investigated: the first specimen is bare cylinder with a scale ratio of 1.200. Diameter, length, and thickness of the sample are 101, 1080, and 2 mm, respectively; the cylinder is kept at a draft of 917 mm, while the bottom of the specimen sealed watertight. The second specimen is a cylinder of the same geometric dimensions but attached with vertical stripping wires. The third specimen is a cylinder with the same dimensions but fitted with helical stripping wires.

Protrusions and surface imperfections on a cylindrical member shall influence flow around the member (Sarpkaya, 1977). Two steel wires, as shown in Fig. 3.10 are attached to the surface of the cylinder. Wires are fixed to a miniature turnbuckle and stretched over the length of the cylinder. Wires are restrained to the required position as it passes through two notches, which are fixed at both ends of the cylinder. Wires are kept under the required tension so that it held against the cylinder sufficiently tight without using glue. The diameter of the stripping wire used is 0.3 mm ($d/D = 0.003$), where this is the ratio of the diameters of the wire and the cylinder. Experiments are repeated by changing the position of the pairs of the stripping wires. Wires are fixed at $\pm\alpha$ with respect to the stagnation point, and the included angle (α) is changed in every test setup. Notches are cut at five different angles, and the stripping wires are stretched over these notches each time. Table 3.2 shows the details of the position of stripping wires for each setup.

Helical strakes are referred to as effective suppression systems for VIV in risers, suspended pipelines, and other slender members. Though helical strakes were found to be adequate to suppress VIV under the wind but

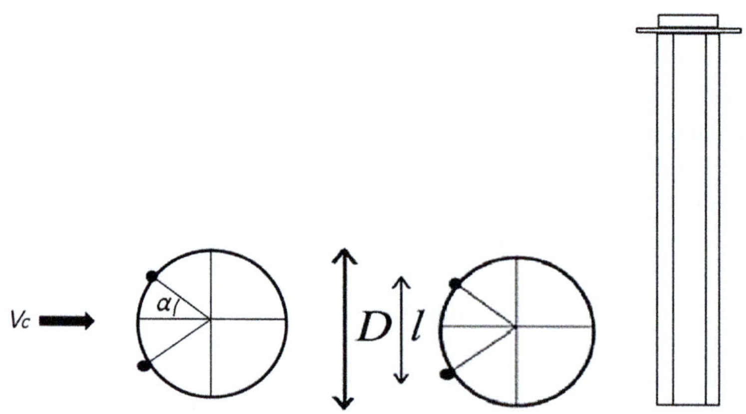

Figure 3.10 Views of cylinder attached with vertical stripping wires.

Table 3.2 Position of wires in different cases.

Description	α	*l/D*
Case 1	40	0.44
Case 2	50	0.55
Case 3	60	0.67
Case 4	70	0.78
Case 5	80	0.89

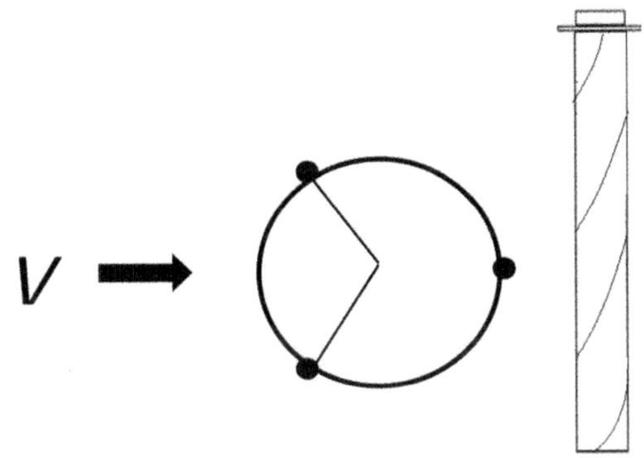

Figure 3.11 Views of cylinder attached with helical wires.

were not effective in the fluid medium. It is because an optimal configuration of helical strakes in liquid medium necessitates a higher pitch. Helical strakes, along with shrouds, are proposed for such applications due to their insensitivity to the direction of the current. For example, SPAR platforms are reinforced with helical strakes to suppress VIM. However, the presence of strakes and shrouds tend to increase the drag on the attached members. Fig. 3.11 shows the view of the cylinder wound with helical wires. Table 3.3 shows the details of different cases, where the cylinder is fitted with helical wires. Wires are placed at 120-degree apart around the surface of the cylinder and wound uniformly along its length. It is to ensure that wires are placed symmetrically in the direction of the current.

3.11 Effect of suppression on vortex-induced vibration

As the fluid past a bare cylinder, which is not deployed with any VIV suppression system, the formation of alternate vortices induced transverse vibration. Empty cylinder vibrated with an expected pattern of shedding

Table 3.3 Cylinder with helical wires.

Experiment no:	Pitch	Length of wire used (m)
Case 1	5D	1.272
Case 2	7D	1.183
Case 3	10D	1.139

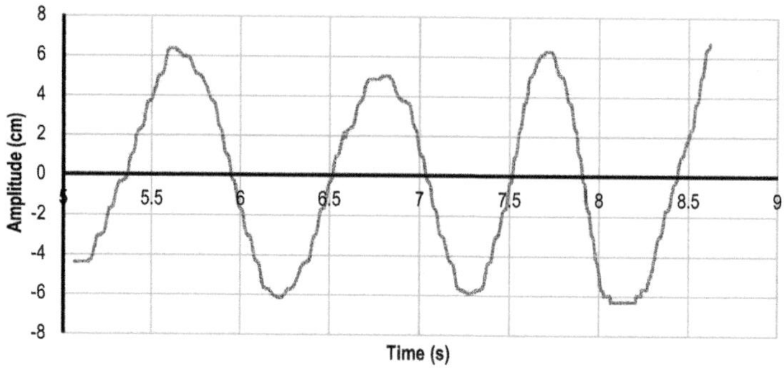

Figure 3.12 Transverse vibration of bare cylinder ($V_{rn} = 5.44$ m/s).

under the action of vortices on either side of the cylinder. Fig. 3.12 shows the vibration time history of the bare cylinder under the current velocity of 0.8 m/s, which amounts to a reduced velocity of 5.44 m/s. The plot shows a periodic vibration, with a zero mean. Under the different range of the reduced velocities, the transverse vibration of the bare cylinder is also plotted, as shown in Fig. 3.13. It is seen that the amplitude ratio increases with the increase in the reduced velocity. At initially reduced velocities the cylinder vibrates with a small amplitude. The shedding frequency of vortices changes with the rise in the velocity. It resonates with the frequency of the bare cylinder. It resulted in a sudden increase in the response amplitude. A lock-in region sets in at a reduced velocity of 6.668 m/s, where the current velocity is about 0.98 m/s. It resulted in the maximum response amplitude of occurred ($A/D = 0.87$). The response is sustained further over a range of reduced velocity, which is followed by a reduction in response.

The obtained responses are compared with those of the amplitude ratios in the literature (Khalak and Williamson, 1991), as shown in Fig. 3.14. The discrepancy in both plots is attributed to the difference in the mass and damping ratios. The current study has a mass and damping ratio as unity and 9.36%, respectively, and is different from that of the referral study.

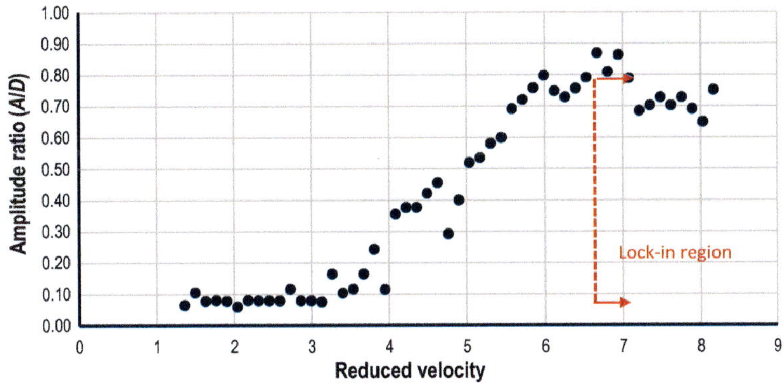

Figure 3.13 Amplitude ratio of bare cylinder for different reduced velocities.

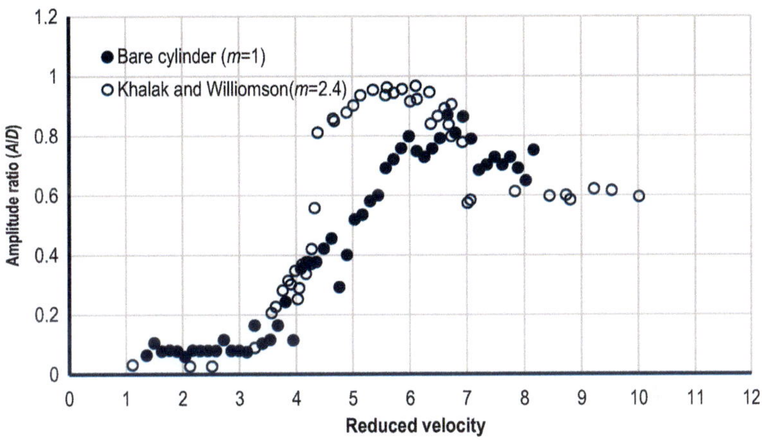

Figure 3.14 Comparison of amplitude ratios of bare cylinder.

A considerable margin of difference in the damping ratio played a crucial role in the variation of results. Fig. 3.15 shows the influence of the reduced velocity on the frequency of oscillation of the bare cylinder, which has no VIV suppression system. A linear relationship ensures the fact that the Strouhal number remained constant during the tests.

The presence of vertical stripping wires modifies the flow due to the angular deflection of flow caused by wires, and the shift of frequency due to structural modifications. Stripping wires are capable of creating turbulence in the boundary layer, resulting in a larger angle of separation along the circumference of the cylinder. It is expected to reduce the width of the

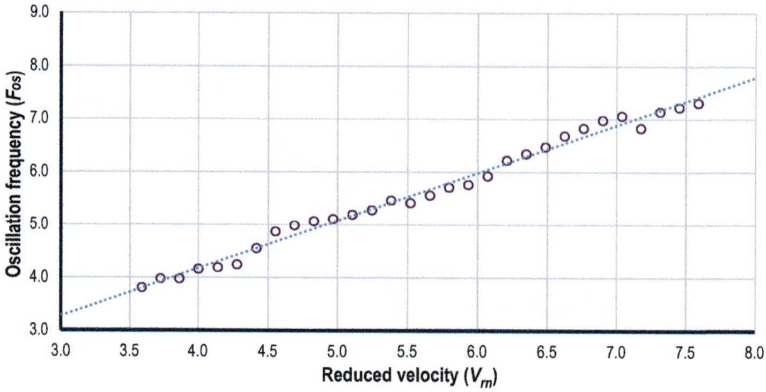

Figure 3.15 Influence of reduced velocity on the oscillation frequency of bare cylinder.

wake formation in addition to lowering the alternating forces. Flow breaks at the point of attachment when a pair of vertical stripping wires are attached to the surface of the cylinder. For the current configuration, it is to be noted that it is a unidirectional suppression configuration in which the flow encounters the stripping wire at a constant angle of approach. Tests are also carried out by placing the stripping wires at different angles ranging from 40 to 80 degrees. Fig. 3.16 shows the variation of amplitude ratio of the cylinder fitted with stripping wires at different angles.

It is seen from the figure that the amplitude ratio of the cylinder with striping wires fixed at 50 degrees is found to be very useful in the chosen range of reduced velocities. In comparison to the bare cylinder the reduction of amplitude ratio is about 72%, which is quite significant. It is also observed that the striping wires, placed at an included angle of 40 degrees, do not offer effective suppression. It is also the same as the angle is further increased from 50 degrees onwards. It is interesting to note that attaching stripping wires, irrespective of the included angle between their pairs, controlled the vortex shedding effects, by causing a reduction in wake formation. Under steady flow conditions the change in the position of the attachment of wires influenced the behavior of the cylinder. A notable improvement in suppression is observed when wires are placed at angle 50 and 60 degrees from the stagnation point. In the chosen range of reduced velocity a clear lock-in is not seen in both cases. In other instances the lock-in point is shifted to the higher end of reduced velocity as the position of wire is widened from stagnation point. Table 3.4 shows the summary of the comparison.

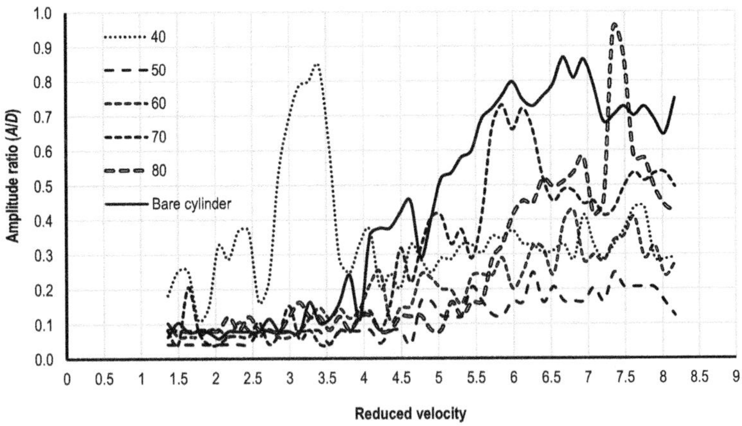

Figure 3.16 Amplitude ratio of cylinder with vertical stripping wires at different angles.

It is seen from the table that the value of peak amplitude ratio occurred is lesser than half of that obtained in a bare cylinder. Adopting a 40-degree arrangement provided a slight improvement in performance at lower reduced velocity and a notable increase in higher decreasing velocities.

From the set of studies discussed previously, it is seen that attaching vertical stripping wires is effective in suppressing the VIVs. But the performance of the cylinder improves significantly with the change in the angle of flow to the position of wires.

Another set of experiments is carried out by attaching helical wires, to make the VIV suppression system, direction insensitive. Unlike strakes, helical wires can be mounted without causing a large projected area, which would be normal to the free-stream flow. Hallam et al. (1977) suggested fins as an effective suppression. The width of the fins shall be about 10% of that of the diameter of the cylinder. The pitch shall be about five times that of the diameter of the cylinder. As seen from the earlier set of experiments, stripping wires with 50- to 60-degree angles are found to be effective. Hence, helical stripping wires with an included angle of 120 degrees are attempted as this would result in a symmetric arrangement. One of the parameters altered in this investigation is the pitch of the helix; experiments are carried out for three different angles of the helix, namely, 5D, 7D, and 10D. Fig. 3.17 shows the response of the cylinder fitted with helical stripping wires at a different pitch. Helix with 5D showed no effective suppression. For all the three cases considered in the

Table 3.4 Comparison of vortex-induced vibration suppression using vertical wires.

Description	Bare cylinder	40 degrees	50 degrees	60 degrees	70 degrees	80 degrees
Area under the curve	2.970	2.378	0.818	1.356	2.069	1.79
Maximum amplitude ratio	0.8685	0.8454	0.2478	0.4259	0.7304	0.9499

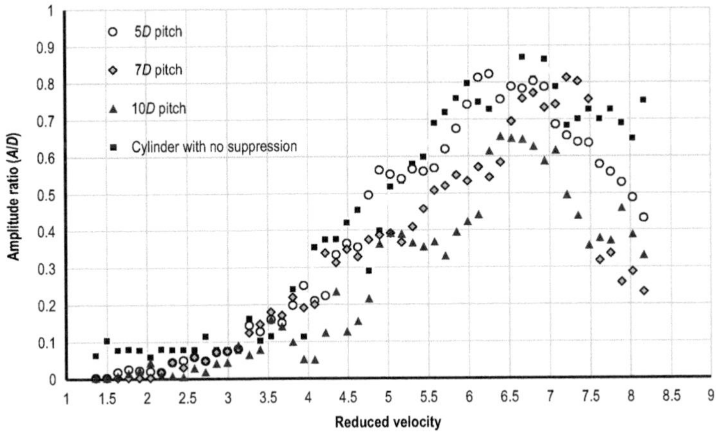

Figure 3.17 Response of cylinder attached with helical wires at different pitch of the helix.

study, VIV suppression shows a similar trend, except that there is a marginal variation in the peak amplitude ratios and the velocity at which lock-in occurs (see Fig. 3.16). In the case of the cylinder fitted with helical stripping wires, flow encounters the strip wires along the span of the cylinder at different angles to the stagnation point. Marginal improvement is seen in suppressing the amplitude ratio for a helix with varying values of the pitch. One of the benefits of helical wires is that it is omnidirectional compared to that of the vertical stripping wires. It has a practical application where the direction of flow is not constant. In a helical arrangement, wires wrapped at a pitch of about 10 times the diameter of the cylinder showed the most effective suppression. In this case the response amplitude is reduced by about 25% with that of the bare cylinder. The effectiveness of each arrangement is studied by considering the area under the curve, which is presented in Table 3.5.

Table 3.5 Comparison of vortex-induced vibration suppression using helical wires.

Description	Bare cylinder	5D pitch	7D pitch	10D pitch
Area under the curve	2.970	2.6791	2.2728	1.807
Maximum amplitude ratio	0.8685	0.8221	0.8124	0.6538

As seen from the table, the area under the curve is lesser than that of the bare cylinder, indicating that helical stripping wires are capable of suppressing the vortex shedding effectively. The area is reduced by about 10%, 23%, and 40% in the case of the pitch of the helix at 5D, 7D, and 10D, respectively. Effective reduction in these cases is because helical stripping wires offer VIV suppression, which is insensitive to the direction of flow. Besides, when wires arranged helically on the surface of the cylinder, it can break up the correlation of vortices along the length of the cylinder.

VIV is the outcome of exciting forces that are generated by vortex shedding on the hull of a bluff body, causing a response closer to the resonant period. It is gaining much importance in different engineering systems, including offshore structures. This phenomenon can induce fatigue stresses on slender bodies such as risers. It can also result in significant hull motion on rigid bodies, which is unacceptable. Many efforts are expended to understand and eliminate the consequences of VIV to avoid the repairs caused by excessive vibrations. Observation of significant amplitude responses of systems that used water as working fluid necessitates the importance of VIV suppression by design. It is found that various modifications to structural members prone to VIV can alter the flow around the body and thus eliminate consequences from VIV.

Among various methods proposed to suppress VIV, strakes and shrouds are found to be active regardless of the orientation of structure to waves and current but showed a substantial increase in drag, resulting in high drift. When strakes or bumps are fixed, no regular shedding is observed on the wake side of the cylinder. One of the significant advantages is the active suppression of the amplitude of vibration of the cylinder. These designs also encounter some practical difficulties, such as marine growth. Protrusions that are appropriately designed can trip the boundary layers causing a substantial change in the flow around the bluff body in the desired manner. Stripping boundaries will change the separation point of the flow around the body. Properly positioned wires can reduce the wake width and the other forces that arise from vortices. These effects are practical even when the diameter is smaller than the laminar boundary layer thickness.

Under the action of the steady current a bare cylinder undergoes a transverse vibration whose amplitude becomes almost equal to its diameter. Presented discussions on experimental studies confirm the effectiveness of stripping wires in VIV suppression. It is seen that when stripping wires are arranged vertically, the flow separation point has an essential role in the vibration. As the angle between wires with respect to that of the stagnation point increases, the lock-in point is shifted toward the higher reduced velocity. Based on the experimental investigations carried out, it is verified that vertical wires, when placed at 50 and 60 degrees, are found to be effective in VIV suppression. It has a practical application where a prominent current direction prevails. Alternatively, VIV suppression is also attempted using helical configuration. As this configuration is symmetric to every direction, it can provide effective suppression irrespective of the current path. In this case the pitch of the helix is critical. Out of helical configurations tested, wires placed at a pitch equal to 10 times the diameter of the cylinder showed maximum suppression with a reduction of about 25% in peak amplitude, in comparison to that of the bare cylinder.

Exercise

Part A: Objective questions

1. When a flowing medium strikes a nonstreamlined __________ (bluff body), it moves around, causing alternating pressure on either side
2. __________ (Vortex-induced vibration) is a periodic exciting force, creating a response closer to the resonance period of the encountering body.
3. Among various methods proposed to suppress VIV, __________ and ________ (strakes, shrouds) are found to be active.
4. VIV suppression systems cause substantial increase in ________ (drag), resulting in high drift.
5. What are the consequences of VIV?

   It results in problems of twofold: (1) destructive overstress on structural members due to larger amplitudes and (2) strong potential to cause long-term effects such as fatigue failure
6. __________ (boundary layer) is a thin layer confined to the surface of the body.
7. Viscous effects are very prominent inside __________ (boundary layer).
8. Vortices are shed from bluff structures in __________ (subsonic) flow.

9. Briefly explain lock-in.

When the natural period of the member synchronizes with the shedding period, the member will lock-in to vortex shedding frequency, termed lock-in phenomenon.

10. __________ (helical strakes) are deployed to mitigate VIM on smokestacks and risers

11. How is lock-in related to reduced velocity?

Lock-in is related to the reduced velocity and is given by the following expression $V_{rn} = \frac{V_c T_n}{D}$

12. What are loop currents and cold-core eddies?

Loop currents are generally present on the near-surface, whereas maximum cold-core eddies are present at a depth around 400 m below the mean sea level.

13. What are the two approaches to address VIV in risers?

There are primarily two approaches to address the issues related to VIV in risers. Either the design or geometry is modified to eliminate VIV, or appropriate VIV suppression devices are to be used as external attachments to the riser.

14. Name a few VIV suppression devices.

VIV suppression devices include strakes, fairings, shrouds, and dimples. Among them the most commonly used are helical strakes and fairings. The helical strakes consist of fins, which are wound helically around the surface of the riser.

15. Name a few parameters that should be addressed while modeling.

Some of the parameters should be addressed while modeling VIV. They are geometric similitude, hydrodynamic scaling, dynamic similitude, and degrees-of-freedom of the scaled model. Geometric similitude implies the similarity in shape.

Part B: Descriptive questions

1. Write a detailed discussion on vortex-induced vibration, explaining the conditions necessary for its formation, its effects, and methods of controlling VIV.
2. Explain vortex-induced motion.
3. What is a lock-in phenomenon?
4. How Reynolds number influences VIV?
5. Explain the effect of vertical wires and helical strakes in VIV suppression.

CHAPTER 4

Recent advancements in materials and offshore platforms

4.1 Introduction

Different types of materials are used in the construction of ocean structures. They are used for a variety of purposes, namely, installation, repair, rehabilitation, and corrosion protection. A wide range of materials such as composites, concrete, mild steel, copper, aluminum, fiber-reinforced plastics, titanium, and other buoyancy materials are preferred. The choice of their use depends on the application requirements as all the above cannot be used for the same purpose (Chandrasekaran and Jain, 2016). Therefore the selection of materials for the marine environment is a complex task as no single material characteristic shall rank the suitability of the material for offshore applications, in particular. Detailed knowledge about their structural characteristics and relevant code compliance is necessary. In addition to the complexities that arise due to the innovative geometric forms of offshore structures, the combination of several environmental loads such as a wave, wind, current, ice, and impact loads makes the design more complex. The behavior of materials under high temperature and fire is also one of the critical design considerations as offshore structures are designed for a high level of reliability (Chandrasekaran and Gaurav, 2017; Chandrasekaran, 2015a,b, 2016a,b, 2017; Chandrasekaran, 2019a,b). Offshore drilling and production platforms, surface buoys, instrument platforms, and underwater vehicles are a few examples of newly developed ocean systems, the design of which is complicated also due to materials in use. Materials used in these systems require unique characteristics, besides, to be compatible with extreme weather conditions. It is, therefore, essential to note that the structural design of offshore and marine structures is just not another conventional type of design, which is strength dominant (Chandrasekaran and Subrata, 2012; Chandrasekaran, 2014). This process is also dominated by material characteristics in addition to the critical combination of various environmental loads.

Design of Marine Risers with Functionally Graded Materials
DOI: https://doi.org/10.1016/B978-0-12-823537-9.00005-0

4.2 Selection of materials

Under the given choice of a wide variety of materials available for use in offshore and marine structures, a few parameters govern their choice: type of applications, where a specific material is chosen, structural properties under regular, cyclic loads, high and very low temperature, recycling characteristics, nontoxic nature and biofriendly characteristics, sustainability for the service life of the structure under marine environment, and the code compliance. It is interesting to note that unlike other conventional design procedures, which are governed by the codes, the design of marine structures is guided by code requirements. The use of advanced materials is always welcome; even their strength-related characteristics are referent values in a particular model. For example, as seen in the earlier chapter, structural properties of functionally graded material (FGM) may not be referred by any code for its use in a design application but still tried out for performance-based applications and considered as referent design, if found successful. To be very specific the behavior of FGM under cyclic loads is not available for reference in international codes, as the use of FGM in marine applications is in the state of progress. But, still, to assess their fatigue life, a comparative (or conservative) values from the relevant codes are used in the analysis for comparison. It emphasizes a critical approach that the use of materials in offshore and marine structures is recommended after intensive research in the materials lab. Their applicability is also well disseminated in the public domain for detailed commentary and advice. As the marine environment influences the performance of materials significantly, the choice of appropriate material for ocean structures becomes a critical engineering decision.

Members are offshore structures in general, and risers, in particular, are subjected to a variety of loads and their combinations. The form-dominant capability of offshore compliant structures enables an effective load disbursement, but their materials are also submitted to several challenges. In addition to meeting the design requirements, materials should withstand hazards that arise during drilling and production operations, without undergoing catastrophic failure. As risers are under dynamic loading, the fatigue resistance of the chosen material also becomes very vital. Innovative materials and varying shapes of cross sections are being attempted by various researchers in the past (Chandrasekaran and Nagavinothini, 2018a,b, 2019a,b). It is, therefore, necessary to ensure that the chosen material for construction should possess a large availability in the desired shape, size, and modular sections, apart from being cost-effective.

Mechanical properties are generally considered as essential indices to study the behavior of metals under loads. These properties have estimated how the material resists a variety of forces. More focus is on their functional and structural degradation in a corrosive environment such as an ocean. Some of the essential properties are strength, hardness, elasticity, plasticity, brittleness, ductility, and malleability. As we understand that the forces acting on offshore structures induce a three-dimensional stress state, it is more important to decide upon the fact that under which stress state and type of failure, one should classify the ultimate strength. While some materials are equally strong in compression, tension, and shear, a few show a significant difference. By careful alloying, the metals, a combination of plasticity and strength is achieved and used in marine construction. See, for example, the FGM combinations discussed in the second chapter, which also highlights the manufacturing process such as wire-arc additive manufacturing (WAAM).

A combination of various forces acting on risers, in particular, results in a complex behavior of the material. One classic example could be the dynamic tension variations of top-tensioned risers (TTRs), held down from a semisubmersible. These variations can result in fatigue failure in addition to the failure caused by excessive deformation in geometry and shape of the riser. The ductile behavior of the steel is well understood from the stress—strain curve plotted under the uniaxial tensile load. But, their response is different under high temperatures. It is a classic case of importance in marine risers as they operate under high temperature and pressure in the presence of hydrocarbons that are transported.

4.3 Marine environment

Steel is one of the most favorite materials for marine risers, but its behavior at high temperatures makes the design different from those of the conventional ones. Fig. 4.1 shows the variations in material characteristics with the heat. It is observed that vital properties of structural steel such as modulus of elasticity, stiffness, and yield strength decrease with the increase in temperature, while the ductility increases showing an indication of strength development (Chandrasekaran and Nagavinothini, 2019a, b). The other properties of carbon steel under high temperatures also vary significantly and need to be considered in the design. Figs. 4.2, 4.3, and 4.4 show the variations of thermal conductivity, specific heat, and thermal strain of carbon steel with temperature, respectively. It is seen from the

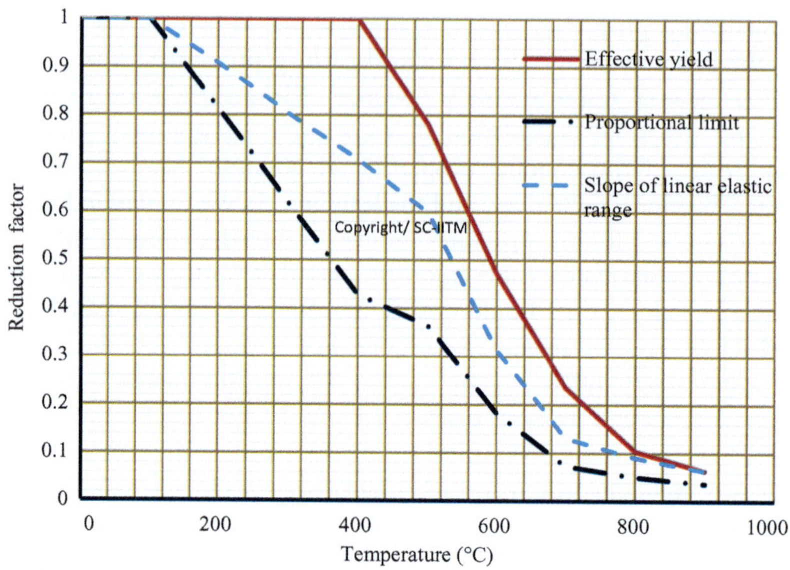

Figure 4.1 Variation in material characteristics of carbon steel with temperature.

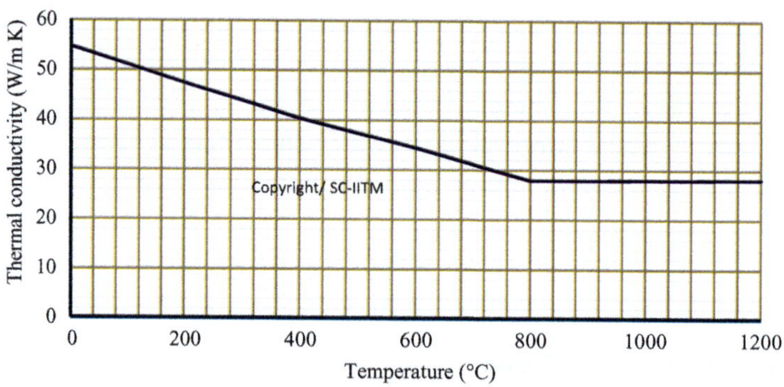

Figure 4.2 Variation of thermal conductivity of carbon steel with temperature.

figures that in case of carbon steel, thermal conductivity drops with the increase in temperature up to about 800°C, which is termed the critical temperature. It is around this temperature range; the specific heat of carbon steel shows a maximum value, while thermal strain variation remains as a constant plateau (Chandrasekaran and Nagavinothini, 2019b).

Although steel is one of the most preferred construction materials for marine risers, strength degradation, and decrystallization of steel at higher temperatures remain a significant challenge. Composites can be thought of as suitable alternatives but mostly recommended for nonstructural

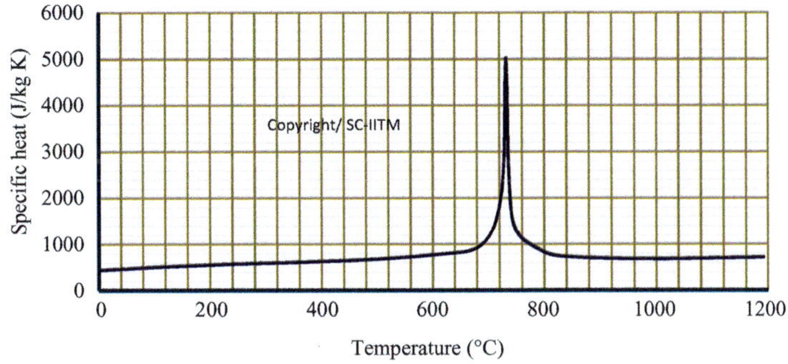

Figure 4.3 Variation of specific heat of carbon steel with temperature.

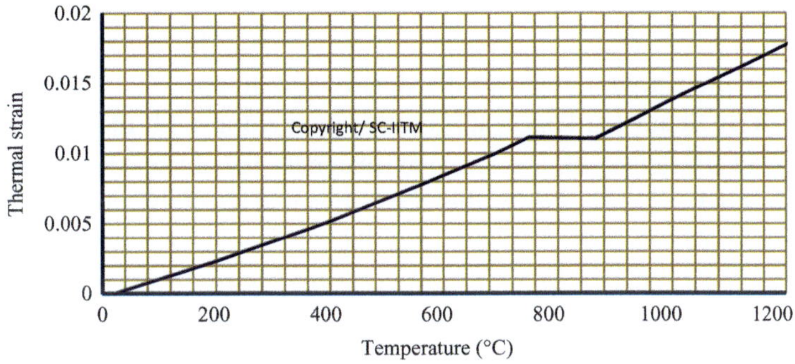

Figure 4.4 Variation of the thermal strain of carbon steel with temperature.

members. Delamination of composites limits their use as an engineering material. Under high temperature and pressure, thermal stresses occur at the bonding interface due to the differences in their coefficients of thermal expansion. It leads to a crack formation and results in delamination at the interface. As seen in the previous chapter, FGM is being attempted for members under high-temperature and high-pressure conditions; a typical application will be marine risers (Chandrasekaran et al., 2019). As marine risers are subjected to a corrosive environment due to the presence of hydrogen sulfide, chloride, and CO_2 gases, it is preferred to use a corrosion-resistant alloy of duplex stainless steel. But, the alloying element should offer both the corrosion resistance and strength as well. Hence, material, which is functionally graded with carbon—manganese steel and nickel as buffer layers along the inner portion of the riser, is chosen as a prima facie research material; Titanium Grade 2 is graded functionally along the external surface of the riser (Chandrasekaran and Jain, 2016). Alternative design techniques to reduce forces on risers are also suggested by the researchers (Chandrasekaran and Merin, 2016; Chandrasekaran and Madhavi, 2014a,b, 2015a,b,c; Chandrasekaran et al., 2015b).

Functional grading combines the advantages of the grouped materials. For example, duplex stainless steel has high corrosion resistance to the marine environment and also possesses higher resistance to stress corrosion cracking (Chandrasekaran and Jain, 2016; Chandrasekaran and Gaurav, 2017). Specific properties such as the yield strength and tangent modulus of the individual materials are technically manipulated to arrive at a functionally graded combination. One of the successful combinations is seen as a functional grading of 22 Cr duplex stainless steel with ASTM Titanium Grade-2 at the internal and external surface, respectively. A buffer layer of nickel is used to enable the manufacturing process of this FGM using a WAAM. A typical trial section of the FGM is shown in Fig. 4.5, while Table 4.1 shows the details of materials.

A bilinear, strain-hardening approach is used along with the yield strength and tangent modulus of the original materials to obtain the stress–strain curve of FGM. Fig. 4.6 shows the comparison of stress–strain curves of the original materials used for manufacturing FGM. It is further recommended as a prima facie material to be used in marine risers. Stress–strain curves of the individual elements are taken from the literature: pure nickel (Det Norske

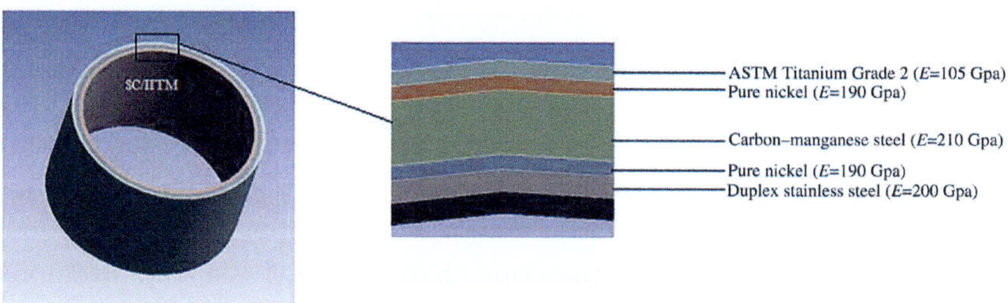

Figure 4.5 Section of the functionally graded riser.

Table 4.1 Details of a functionally graded marine riser material.

Layer	Material	Thickness (mm)
1	Duplex stainless steel (22 Cr) 2209	3
2	Pure nickel buffer layer	2
3	Carbon–manganese steel ER70S6	8.47
4	Pure nickel buffer layer	2
5	ASTM Titanium Grade 2	2
	Total thickness	17.48

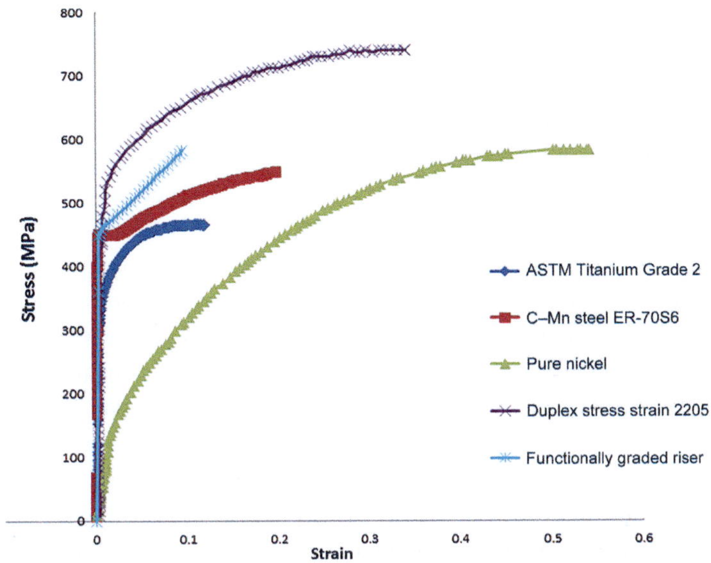

Figure 4.6 Comparison of stress–strain curves of individual materials used for FGM. *FGM*, Functionally graded material.

Table 4.2 Structural properties of different materials used for grading.

Material	Modulus of elasticity (GPa)	Poisson's ratio
Carbon–manganese steel	210	0.30
Pure nickel	190	0.315
Duplex stainless steel	200	0.30
ASTM Titanium Grade 2	105	0.37
Functionally graded combination	174.506	0.2987

Veritas, 2012), ASTM Titanium Grade 2 (Det Norske Veritas, 2002), and carbon–manganese steel (Det Norske Veritas, 2012) and duplex stainless steel (Det Norske Veritas, 2018). Table 4.2 shows the structural properties of the individual materials used for grading. The table clearly shows a comparable value of the yield strength of FGM, in addition to making it corrosion-resistant.

As discussed in the earlier sections, the marine environment poses severe challenges to the offshore structures through the environmental loads they encounter. Besides, it also imposes a threat to the materials used for their construction, leaving the choice of material very limited. It is well established that the new-generation offshore platforms resist the loads not by strength but by undergoing large displacements. It is important to note that only those materials, which possess high ductility, can cater to such special requirements. For example, large deformation induced in the design concept such as tension leg platforms (TLPs) demands high ductility in tether materials. Similarly, articulated towers require more fatigue resistance in the materials used for articulated joints. Further, materials chosen for marine construction should be easy to fabricate and transport to the offshore site, which is also one of the significant constraints in the selection of material. In particular, riser materials impose a parallel challenge of the compact form of transporting and laying the risers conveniently. It is reemphasized that the recommendations of international codes govern the selection of material, but alternatives emerge from the conduct of advanced research; FGM is only such recent advancements. The environmental loads acting on offshore structures in general and marine risers, in particular, are highly random. It is, therefore, necessary that failure of material by its strength even occurs should have a lower probability. The intuited failure should not be sudden, or else it may lead to loss of assets that are invested on the platform. Materials chosen for offshore structures should be capable of withstanding the unforeseen hazards besides meeting the general design requirements (Chandrasekaran and Jain, 2016). In the following sections, alternate materials suitable for marine applications are discussed. These discussions are not in their metallurgical perspective but the structural engineering point-of-view.

4.4 Titanium

The majority of offshore applications use titanium for submarine ball valves, pumps, and heat exchangers. Hulls of deep-sea submersibles are preferable with titanium. Other significant applications include water-jet propulsion systems, propeller shafts, exhaust stack liners, navel armors, underwater manipulators, high-strength fasteners, yacht fittings, and piping systems. Titanium alloys are commonly available in manufactured products and found a full application in the aerospace industries, but scarce in offshore engineering. Titanium possesses a few advantages features to their structural characteristics, apart from being an expensive choice. Titanium is as strong as steel but about 45% lighter. High strength, low density, and corrosion resistance of titanium contribute toward its cost reduction in long-term maintenance even though the initial cost is prohibitively high. Titanium has a melting point (about 1700°C), which is higher than that of steel and, therefore, useful under high-temperature applications. As its specific gravity is 4.5, which is about 55% of that of steel, it has a higher strength to weight ratio. In the case of floating and compliant offshore structures, this is very vital. Also, top tension in TTRs that arise due to the self-weight shall be reduced significantly. As titanium requires no corrosion allowance, risers can be designed to satisfy the minimum requirements for mechanical strength and handling. The hexagonal close-packed crystal structure of titanium makes it more reactive when exposed to the marine environment. Upon exposure, it reacts with atmospheric oxygen to form a protective layer and makes it corrosion-resistant.

The most preferred material for marine risers is a variety of grades of X52 steel, or higher. In the present scenario, drilling activities are carried out in deep and ultradeepwater conditions that are prone to high temperatures and pressure. Further, there is a higher probability of corrosion due to the presence of hydrogen sulfide and chloride gases. Therefore the current focus of the oil and gas industries is on developing a riser material with lesser weight, higher corrosion resistance without compromising on its safety and durability. Metals such as titanium and composites are used as alternative materials for risers. Titanium alloy found its first use in the Heidrun TLP drilling riser system. It was used for the body of the riser pipe, flanges, fasteners, and the booster lines. On comparing the density, titanium possesses only half the density of steel, which in turn resulted in the reduction of weight and tension requirements, dimensions of the flange as well as the elimination of the flex joints.

The titanium also offers a high yield strength that reduces the overall wall thickness and weight. The elastic modulus of titanium is only half as that of the steel, as a result of which there is a reduced stiffness and induced bending moment eliminating the need for a flex joint. Another advantage is in terms of durability and resistance to corrosion. Titanium exhibits fatigue property better than steel and also has a better endurance limit. Titanium undergoes an allotropic transformation from α- to β-Ti at 882°C while alloying elements influence this transformation. Manipulation of these crystallographic variations through thermomechanical processing leads to the development of a wide range of titanium alloys. Titanium alloys have good fatigue resistance and high fracture toughness apart from being nonmagnetic. It also possesses a good heat transfer property with low heat thermal conductivity, low coefficient of thermal expansion (about 9–11 ppm/°C), and nontoxic, which makes it biologically compatible. It is also evident that inflowing or static seawater at temperatures up to 130°C, titanium surfaces are immune to corrosion while other metals and alloys corrode significantly. Titanium is immune to crevice corrosion up to at least 70°C in seawater, while steel tends to corrode even at 10°C.

4.4.1 Classifications and alloying elements

Titanium alloys are classified into three groups: alpha, alpha–beta, and beta; classification depends on its microstructure. Depending on the type and amount of alloying additions, the transformation temperature of α- to β-Ti can be raised or lowered. The addition of alloying elements classifies the equilibrium transformation into two temperatures: α-transus and β-transus. In a broader sense, titanium alloys are classified based on the phase diagrams: Types I–III. Phase diagram includes a line to indicate the temperature at which β begins to undergo diffusionless transformation to a supersaturated α. Type I shows greater solubility of additional elements in α than in β, whereas in Type II, it is reversed. Additional elements stabilize in the appropriate group while the remaining is elevated to a higher temperature by increasing the concentration of the alloying elements. Alpha phase is stabilized by elements with an electron-to-atom ratio, lesser than four; for example, oxygen, carbon, nitrogen, and aluminum. The beta phase is stabilized by elements with an electron-to-atom ratio greater than four; see, for example, iron, molybdenum, and manganese. Elements such as zinc, tin, and silicon have an electron-to-atom ratio equal to four and are considered to be neutral.

Pure titanium has low strength and high ductility. The addition of small amounts of elements to their chemical composition increases their strength and decreases ductility. Properties of titanium can be altered with the alloying elements that are compatible with titanium. For example, alloying aluminum with titanium increases a few structural properties such as tensile strength, creep strength, and elastic modulus. But, if increased more than 6% by weight, the formation of an intermetallic compound (α_2) induces brittleness. Tin, if alloyed with titanium, dissolves in both alpha and beta as well. It is a weak alpha stabilizer that results in a solid solution and hardens the alloy with aluminum, without causing embrittlement. Zirconium is alloyed with titanium, retards the rate of transformation as it is a weak beta stabilizer. If added more than 6% by weight, it reduces ductility and creep strength. Molybdenum is a strong beta stabilizer and, if alloyed, helps to increase hardness but reduces longtime, high-temperature strength. It also reduces the weldability significantly. Alloying niobium improves the resistance to high-temperature oxidation as it is a strong beta stabilizer. Iron, if alloyed with titanium, reduces the creep strength. Carbon, being a strong alpha stabilizer, widens the temperature range between alpha and beta transitions.

Table 4.3 compares different vital properties of titanium with that of other metals and alloys (Chandrasekaran and Jain, 2016). It is seen that

Table 4.3 Titanium: a comparison.

Mode of corrosion	Copper-based alloy	Stainless steel 316	Stainless steel 6 Mo and duplex	Titanium alloys
General corrosion	Resistant	Resistant	Resistant	Resistant
Crevice corrosion	Susceptible	Susceptible	Susceptible (>25°C)	Resistant up to 80°C
Pitting corrosion	Susceptible	Susceptible	Resistant	Immune
Stress corrosion	Susceptible	Susceptible (>60°C)	Resistant	Resistant
Corrosion fatigue	Susceptible	Susceptible	Susceptible	Immune
Galvanic corrosion	Susceptible	Susceptible	Resistant	Immune
Microbiological corrosion	Susceptible	Susceptible	Susceptible	Immune
Weld/heat-affected zone corrosion	Susceptible	Susceptible	Susceptible	Resistant
Erosion corrosion	Susceptible	Susceptible	Resistant	Highly resistant

titanium alloys are almost resistant to all types of corrosion that are common in the marine environment. Titanium has an evident edge as a construction material for marine applications. The only demerit is the increase in the initial investment, but by including the cost of maintenance over a service life of about 15 years, titanium and aluminum shall become a competitive, affordable, and replaceable material in comparison to steel.

Titanium is heat-treated to improve its fracture toughness, fatigue strength, and high-temperature strength. It also helps in reducing the residual stresses that are developed during fabrication. While annealing helps to improve ductility, machinability, solution treating and aging increase its strength. In the case of hot working with titanium, absorption of certain gases such as hydrogen (above 148°C), oxygen (above 700°C), and nitrogen (above 815°C) causes embrittlement. Therefore to avoid hydrogen absorption, hot working is carried out in an oxidizing atmosphere while the oxygen-embrittled layer can be removed after hot working.

4.5 Composites

Composite risers were first used in Heidrun TLP in the year 2001. A full-scale composite drilling riser joint was used to drill three wells located at different positions along the riser string. A computer-controlled filament winding technique was used for the fabrication of the riser joints. Forty layers of carbon fiber and carbon/glass fiber were applied in the radial and longitudinal directions. It is interesting to note that the composite riser was attempted on the Heidrun TLP after the successful installation of the titanium riser.

Composites consist of two or more constituents, combined in such a manner that they keep their solid phases and are not soluble in each other. They also do not result in the formation of any new chemical compound. One of the constituents is termed the reinforcing phase. The other one in which the reinforcing phase is embedded is called the matrix. Based on the geometry of the reinforcing phase, they are classified as particulate-, flake-, and fiber-reinforced composites. Fiber-reinforced composites are further grouped into continuous fiber, short fiber, and whiskers. Based on the type of matrices, they are classified as a polymer matrix, metal matrix, carbon fiber matrix, fiber-reinforced polymeric composites, and particulate-reinforced metal matrix composites.

There exist hybrid varieties of composites that use multiple reinforcements and matrices. Typical examples are carbon and glass fiber in the epoxy matrix. Glass-reinforced epoxy (GRE) composites are extensively used in the piping systems as possessing a good resistance against corrosive fluids. Pultruded glass or phenolic gratings are the particular type of GRE being commonly used in such process industries. The performance of a composite depends on the process of its fabrication. Pultruded products, due to their high fiber-to-resin ratio (70:30), possess a higher load-bearing capacity. Deck gratings of longer spans are designed using the pultruded gratings as they deflect lesser than that of the molded gratings. It also finds a suitable application in the decks of the topside of offshore platforms. The modular dimensions of the floor are fabricated using the pultruded grating panels to enable prefabricated construction in offshore topside applications. They can be easily cut and modified for the required dimension, shape, and size. Alternatively, phenolic gratings are also preferred for the topside application due to their better performance under fire. They can retain a significant level of functionality even after fire exposure apart from having a low-smoke emission.

Smoke and toxic-resistance, resistance to shock and impact loads make composites as smart alternates for topside in the offshore platforms. The structural and mechanical characteristics of the composites can be altered either by the method of manufacturing or by fabrication. They can be customized to suit special features such as low thermal conductivity, thermal coefficient-of-expansion, higher axial strength, and stiffness. Composites are used in the process lines of any production unit that handles hydrocarbon. One of the main advantages is the high cost of replacing steel pipes in retrofit applications. Heavy metal pipelines are replaced with lighter composites, which also resulted in cost reduction. Besides, composite pipes also used for fire-fighting mains, seawater intake systems, cooling towers, draining systems, and sewerage systems. Due to high cost–benefits, composites replace corrosion-resistant metals such as copper–nickel alloys and titanium.

One of the favorite nonmetallic materials for offshore applications is fiberglass that is reinforced with plastic. They are one of the varieties of composites, which consist of plastic fibers as reinforcing materials. Composites consist of glass fibers, carbon graphite, nylon, silica, or metals are such as steel, aluminum, boron, and tungsten as reinforcing material. Bonding materials are typically epoxies, polyesters, phenolic, and silicon. The most common is the glass fiber with an epoxy or polyester binder. One of the distinct advantages is that it is maintenance-free and highly durable under a variety of operating conditions. But, the major concerns are a significant change in its internal damping and strength reduction when exposed to high ultraviolet rays. It reaches its fatigue strength in 10 million cycles, making the ratio of fatigue-to-tensile strength lesser than 0.25. It is a major concern for engineering applications in the marine environment. Under the marine environment, fiberglass loses strength by the absorption of water when immersed over a long period. While water absorption reduces compressive strength, exposure to ultraviolet rays causes brittleness as it gets laminated upon the application of heat.

4.6 Nonferrous metals and other materials

Nonferrous metals such as cupronickel alloy, which is a mixture of copper and nickel (70:30), is used for condenser applications such as tubes, tube sheets, and pipelines. K-Monel alloy, which is a combination of nickel and copper (65:30), is another alternative. Bronze, which is an alloy of copper and zinc (90:10), finds increasing applications in the offshore industry. It is reemphasized that unlike other structures, materials for offshore applications are selected not only based on their strength but also the performance criteria. One of the oldest materials used in the marine environment is wood. Currently, wood is extensively used for pilings, docks, and similar applications. Flammability characteristics, nonavailability in large size, and deterioration of strength in marine conditions are a few of the serious limitations. Another drawback is its significant change in strength characteristics to the grain orientation. It shows a significant difference when the load is applied parallel to the grains in comparison to that of the perpendicular direction. Glass-reinforced plastics (GRP) are used in fabricating lifeboats and rescue vessels. They substitute for evacuating personnel on board of an offshore platform, in case of an emergency. They are parked in the boat-landing deck and used in case of emergency. GRP, manufactured with fire-retardant resins, are used as survival crafts and are capable of withstanding flames of fire even at 1200°C.

4.7 Buoyancy materials and syntactic foams

Buoyancy materials have specific gravity considerably lesser than water. The most common are wood and gasoline, the specific gravities of which are about 0.5 and 0.7, respectively. Such materials should possess no-water absorption, no-distortion under compression. Few common applications are small submarines, oil well drill pipe, and deep-sea buoys. Syntactic foams are hollow, glass spheres, dispersed in a plastic matrix. Most efficient syntactic foams use glass spheres of microscopic with an epoxy resin binder. They have a high compressive and shear strength with low water absorption. They can be easily handled with woodworking tools. Syntactic foams, which are free from corrosion, are composite materials synthesized by filling a metal, polymer, or ceramic matrix with hollow particles, termed micro-balloons. The presence of hollow particles results in lower density and higher strength. For example, syntactic foam for 600-bar pressure possesses strength about three times higher than the concrete of 80 MPa. When used as reinforcements in reinforced cement concrete, they can achieve both strength and buoyancy. Table 4.4 shows the structural properties of syntactic foams with epoxy resin (Chandrasekaran and Jain, 2016).

Table 4.4 Properties of syntactic foam with epoxy resin.

Properties	Value
Young's modulus	2.1 GPa
Density	640 kg/m^3
Microsphere density	349 kg/m^3
Resin density	1120 kg/m^3
Hardener density	1050 kg/m^3

4.8 Semisubmersible with postulated mooring failure

4.8.1 Semisubmersible

Semisubmersible is one of the floating offshore platforms, preferred over other alternatives due to a few advantages: It has better stability in the harsh environment, a larger deck area, superior constructional and installation features, and higher mobility. The natural frequencies of semisubmersible vary inversely with the draft and length of the platform (Sunil and Mukhopadhyay, 1995; Stansberg, 2008). Hence, the appropriate choice of the geometric shape is an essential criterion in the design of semisubmersible. There exist six degrees-of-freedom for a floating body in space. They are translations along the three principal axes: surge (along the *X*-axis), sway (along the *Y*-axis), and heave (along the *Z*-axis). There also exist rotations about the three principal axes: roll (about *X*-axis), pitch (about *Y*-axis), and yaw (about *Z*-axis). Out of this six degrees-of-freedom, semisubmersibles are designed to remain very flexible in the horizontal plane. It means that a significant surge, sway, and yaw displacements are enabled. To ensure comfortable operations and the required connectivity between the risers and the platform, they are intended to remain stiff in the vertical plane. Hence, heave, pitch, and roll motions are highly controlled. Semisubmersibles are position restrained by mooring lines of different layouts. But, the station-keeping of semisubmersible should ensure compliancy (flexibility) in the horizontal plane. Therefore large displacements are expected in the surge, sway, and yaw motions, which are also undesirable. Hence, mooring lines are attached with submerged buoys at a specific section along their length. Numerical studies showed that there is a maximum reduction in the tension of mooring lines at the location of the buoy, contributing to the maximum buoyancy (Mavrakos et al., 1996; Mavrakos and Chatjigeorgiou, 1997).

The preliminary design of the GVA-4000 semisubmersible showed that pontoon and column members are found to be safe even under harsh environmental loads (Lee et al., 2005). Alternatively, hybrid mooring lines with a spread mooring lead to loss of stability due to dynamic tension variations (Garza-Rios et al., 2000; Odijie et al., 2017). Semisubmersibles are deployed with steel catenary risers to achieve a lesser offset. These risers experience increased damping under the action of waves and current (Kim et al., 2011). Besides, the layout of the mooring system can also influence the motion characteristics of semisubmersibles. Among all possible mooring layouts, spread—catenary mooring, with and without a buoy, is preferred as it does not change slow-drift motion (Hassan et al., 2009; Jang et al., 2019). Different sag-to-span ratios and the angle of inclination of the mooring lines, under various current velocities, also affect the motion response of floating offshore platforms (Wang et al., 2018). On the other hand, deploying buoys in the mooring system reduces stress in the mooring lines and found to be effective in deepwater deployment (Xu et al., 2018).

Buoys, if provided at various sections along the steel moorings, found to increase their fatigue life (Yan et al., 2018). The fatigue life of the mooring line is found to be sensitive to the stress-concentration factor at the weld sections in the mooring lines. The harmonic balance method is capable of handling the nonlinearities that arise both from the periodic waves and restoring forces of the mooring (Umar et al., 2004). An example study of numerical analysis of a semisubmersible, pegged with a new configuration of spread-mooring system with a submerged buoy, is presented. The motion of the platform induces dynamic tension variation in the steel moorings, leading to fatigue damage. The fatigue life is estimated using the S–N curve approach for the postulated failure of the mooring lines.

A semisubmersible platform chosen for the study is shown in Fig. 4.7. It consists of two rectangular pontoons, submerged to provide sufficient buoyancy. Four columns, equidistant both in the longitudinal and transverse

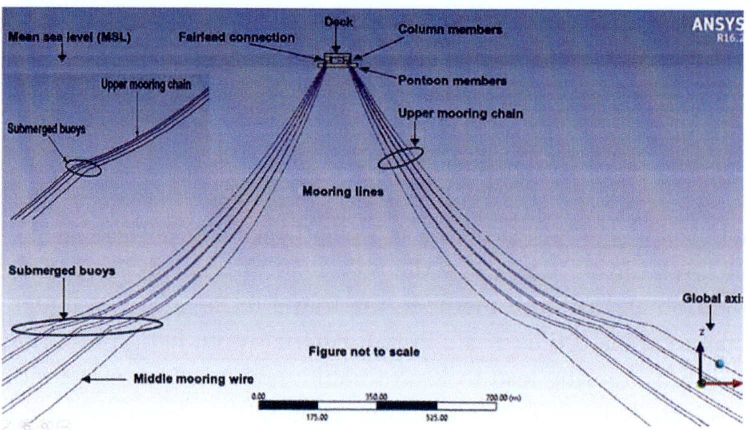

Figure 4.7 Numerical model of semisubmersible.

Table 4.5 Description of semisubmersible.

Description	Value
Water depth	1500 m, 2000 m
Deck size	$74.42 \times 74.42 \times 8.60\ m^3$
Displacement	48,206,800 kg
Pontoon members	$114.07 \times 20.12 \times 8.54\ m^3$
Column members	$17.385 \times 17.385 \times 21.46\ m^3$
Draft	−19.00 m
Freeboard	19.60 m
Metacentric height	16.03 m
Center of gravity	−8.90 m
Radius of gyration (r_x, r_y, r_z)	32.40, 34.10, 34.40 m

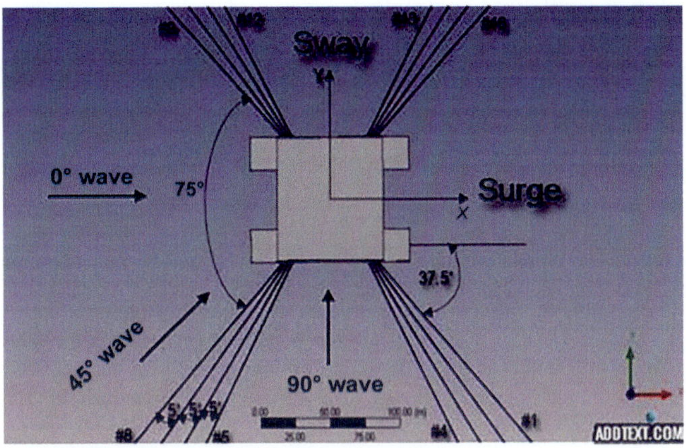

Figure 4.8 Arrangement of the mooring (plan view).

directions, connect the pontoon and support the deck. Four horizontal braces attach the columns to provide the required lateral stability. The floor contains all types of equipment related to the drilling activities, storage of accessories, living quarters, production housing, and a helipad. The description of semisubmersible is shown in Table 4.5.

A spread–catenary mooring, which is symmetric in configuration, is employed for position restraining the semisubmersible. The angle between mooring lines within a group is about 5 degrees, while between each group is about 75 degrees, as shown in Fig. 4.8. Configuration and properties of the spread-mooring system are shown in Tables 4.6 and 4.7, respectively. A submerged buoy is attached at the end of the middle wire in the spread-mooring system, the details of which are given in Table 4.8.

Table 4.6 Configuration of the spread mooring.

Water depth	Length of the spread–catenary mooring		
	Upper chain	Middle wire	Lower chain
1500 m	300	2000	1500
2000 m	500	3000	1500

Table 4.7 Properties of the spread-mooring system.

Description	Upper chain	Middle wire	Lower chain
Mass per unit length (kg/m)	163.7	36.38	163.7
Stiffness (MN/m)	676.81	833.91	676.81
Equivalent diameter (m)	0.095	0.095	0.095
Longitudinal drag coefficient	1.15	0.025	1.15
Transverse drag coefficient	2.4	1.6	2.4
Added mass coefficient	1.0	1.0	1.0

Table 4.8 Description of the submerged buoy.

Description	Value
Diameter	5 m
Structural mass	5861.6 kg
Displaced mass of water	67,086 kg
Added mass	33,543 kg
Coefficient of drag (C_d)	0.3
Coefficient of inertia (C_m)	0.5

The semisubmersible is subjected to environmental loads that arise in the sea state prevailing in the South China Sea. Wave heading angles considered for the study are (0, 45, 90 degrees). API wind spectrum is used to define wind loads, while JONSWAP (Joint Offshore Sea Wave Project) spectrum is used for characterizing the wave loads. The current force is considered as nonlinearly varying for a water depth of 1500 m. The return periods of various events discussed in the analysis are summarized in Table 4.9.

4.8.2 Numerical analyses

The effect of wave loading on the platform is evaluated using the diffraction theory with the boundary element method. JONSWAP spectrum is

Table 4.9 Environmental loads.

Description	Return period	
	10 years	**100 years**
Significant wave height (m)	11.1	13.3
Peak spectral period (s)	13.6	15.5
Peakedness parameter (Υ)	5	7
Wind speed (m/s)	48.3	55
Current speed (m/s)	1.7	1.97

used to include the impact of an irregular wave with the slow drift of the platform. The following expression gives the spectral ordinate:

$$S(\omega) = \frac{\alpha g^2 \gamma^a}{\omega^5} \times e^{\left(-\frac{5}{4} \times \left(\frac{\omega_p}{\omega}\right)^4\right)} \tag{4.1}$$

$$a = e^{\left(-\frac{(\omega-\omega_p)^2}{2\sigma^2 \omega_p{}^2}\right)} \tag{4.2}$$

$$\sigma = \begin{cases} 0.09 \ \text{for} \ \omega > \omega_p \\ 0.07 \ \text{for} \ \omega \leq \omega_p \end{cases} \tag{4.3}$$

$$\alpha = \frac{H_s^2}{16 \int_0^\infty \left(g^2 \Upsilon^a / \omega^5\right) \times \exp\left(\left(-5/4\right) \times \left(\omega_p/\omega\right)^4\right) d\omega} \tag{4.4}$$

where ω is the wave frequency (rad/s), ω_p is the peak frequency (rad/s), g is the acceleration due to gravity (m^2/s), Υ is the peakedness parameter or peak enhancement factor, α is the spectral energy constant, and σ is the spectral parameter. API wind spectrum is used to characterize the dynamic effects of wind loads and is given by the following relationship:

$$S(f) = \frac{\sigma^2(z)}{f} S(\tilde{f}) \tag{4.5}$$

$$S(\tilde{f}) = \frac{\tilde{f}}{\left(1 + 1.5\tilde{f}\right)^{5/3}} \tag{4.6}$$

$$\tilde{f} = \frac{f}{f_p} \ \text{and} \ f_p = 0.025\left(\frac{V_Z}{Z}\right) \tag{4.7}$$

$$\overline{V}_z = \overline{V}_{10}\left(\frac{Z}{10}\right)^{0.125} \tag{4.8}$$

where $\tilde{f}$ is the nondimensional frequency, f and f_p are frequencies (Hz), and $\overline{V}_z$ is the mean 1-h wind speed (m/s).

A nonlinear time-domain approach is considered to assess the coupled response of the platform and mooring lines at every instant of time. The following relationship gives the equation of motion for the analysis:

$$[M + M_a]\ddot{x} + [C]\dot{x} + [K]x = F_{\text{Environment}} + F_{\text{Radiation}} + F_{\text{Mooring lines}} \quad (4.9)$$

where $[M]_{6\times6}$ is a mass matrix, including added mass; $[C]_{6\times6}$ indicates the damping matrix; $[K]_{6\times6}$ is the stiffness matrix, including hydrostatic stiffness; $\{F_{\text{Enviromental}}\}_{6\times1}$ is the external load arising from wind, wave, and current; $\{F_{\text{Mooring}}\}_{6\times1}$ is the load due to the spread–catenary mooring; $\{F_{\text{Radiation}}\}_{6\times1}$ is the load due to radiation force; and $\{x,\ \dot{x},\ \text{and}\ddot{\text{x}}\}$ is the displacement, velocity, and acceleration of the semisubmersible, respectively. The natural periods and damping ratios of the platform at both the water depths (1500, 2000 m), with and without a submerged buoy, are obtained from the free-oscillation tests. The results are summarized in Table 4.10. It is seen from the table that long periods in the surge and sway motion indicate higher flexibility about the horizontal plane, which is a desired feature of a semisubmersible. On the contrary, lesser periods in roll, pitch, and heave show that the platform is stiff in the vertical plane. It is important to note that a semisubmersible is usually characterized by possessing free modes, indicating that the natural periods in all degrees-of-freedom are above the wave periods (Det Norske Veritas, 2016).

4.8.3 Postulated failure

Coupled dynamic response of the semisubmersible is carried out in 1500 m water depth by considering both the cases: mooring lines are intact, and a postulated failure is induced in the mooring lines, deliberately. The latter is an exercise to assess the service life of the platform even under a postulated failure condition of the mooring lines. As mooring lines are generally not designed to fail, a postulated (hypothetical failure) is assumed and intended as a postulated failure. It is important to note that all the mooring lines are not subjected to the postulated failure. Only a few are considered as failed while other mooring lines are intact. It is obvious to note that under the postulated failure condition of a set of mooring lines, adjacent lines share the mooring load. Therefore the platform is left under a free-floating circumstance at any time during the period of the analysis.

Table 4.10 Dynamic characteristics of moored semisubmersible.

Description			Surge	Sway	Heave	Roll	Pitch	Yaw
Without buoy	1500 m water depth	T_n (s)	209	165	21	24	25	49
		ξ_n (%)	6.15	6.92	2.47	2.97	0.93	6.51
	2000 m water depth	T_n (s)	193	184	25	25	25	54
		ξ_n (%)	5.84	6.38	4.15	4.75	1.91	7.98
With buoy	1500 m water depth	T_n (s)	195	183	21	23	24	54
		ξ_n (%)	6.56	6.77	1.20	1.44	1.32	10.36
	2000 m water depth	T_n (s)	213	199	21	23	24	53
		ξ_n (%)	6.87	6.21	1.45	1.06	0.83	4.61

Figs. 4.9–4.11 show the power spectral density plots of the surge, heave, and pitch responses under a 10-year return period for different wave incidence angles, respectively. It is seen from the figures that surge response is significantly higher in the case of postulated failure of mooring

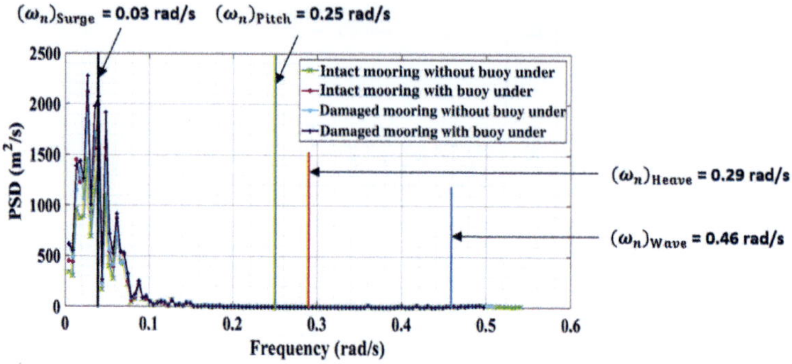

Figure 4.9 Surge response for a 10-year return period (0 degree, 1500 m).

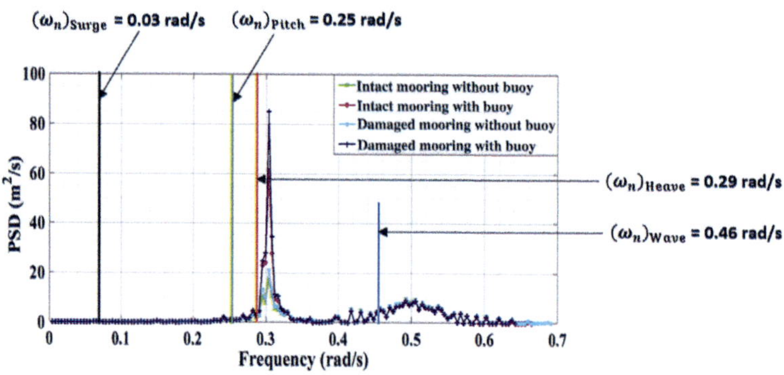

Figure 4.10 Heave response for a 10-year return period (45 degrees, 1500 m).

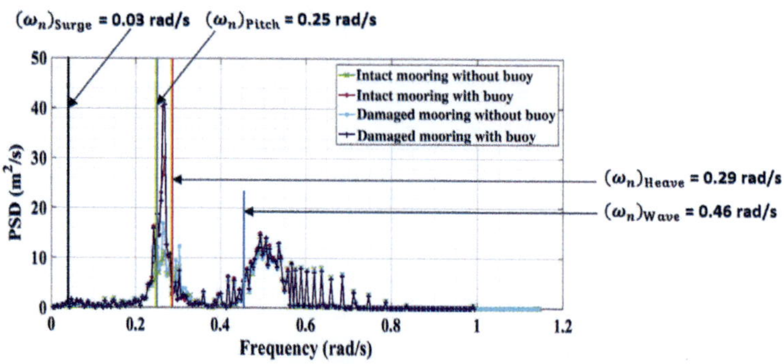

Figure 4.11 Pitch response for a 10-year return period (0 degree, 1500 m).

lines. Also, it could be observed that among all the cases considered in the analysis, the intact mooring without buoy has the least response. Further, the presence of submerged buoy marginally increases the response under 0-degree wave heading. Surge response with intact mooring, pegged with a submerged buoy, is found to be higher. Under the postulated damage conditions, surge response without buoy is significant. It is ascertained to the additional damping of the buoy and higher added mass, resulted from the postulated failure condition. The standard deviation of heave responses is observed to be the highest among all the degrees-of-freedom during the postulated failure condition, for the mooring lines pegged with a submerged buoy. However, in the absence of the buoy, heave responses between intact and postulated failure conditions are insignificant. The maximum pitch response is about 7 degrees under the postulated failure condition. It shows a marginal increase in the presence of buoys. However, this response is well below the permitted limits recommended by the code.

It is also seen from the figures that the natural frequencies of heave and pitch are strongly coupled but, is placed away from the dominant wave frequency, shall avoid resonating response. A few peaks are seen at the closer proximity to that of the surge and heave. They arise due to the impact of the irregular waves with slow-drift response. Such effects are more in the pitch response.

4.8.4 Tension variation under postulated failure

Postulated failure is a set of mooring lines that induce dynamic tension variations in the remaining mooring lines due to load sharing. All mooring lines are pegged with a buoy to assess the reduced response. Figs. 4.12–4.15 show the dynamic tension variations of a set of mooring lines for a 10-year return period, and 1500 m water depth. It can be seen from the figures that the tension variation in mooring line 11 under the postulated failure of (10, 12) is increased by about three times that of its original tension. It is due to the additional load transferred from the adjacent mooring lines that are considered to be failed. Similarly, for a 45-degree wave heading, mooring lines 9 and 10 are under the postulated failure conditions, causing tension variation in the lines 11 and 12, which are adjacent.

Under the increased water depth of 2000 m, as seen in Fig. 4.16, tension variation in the mooring lines (1, 14) has significantly increased due

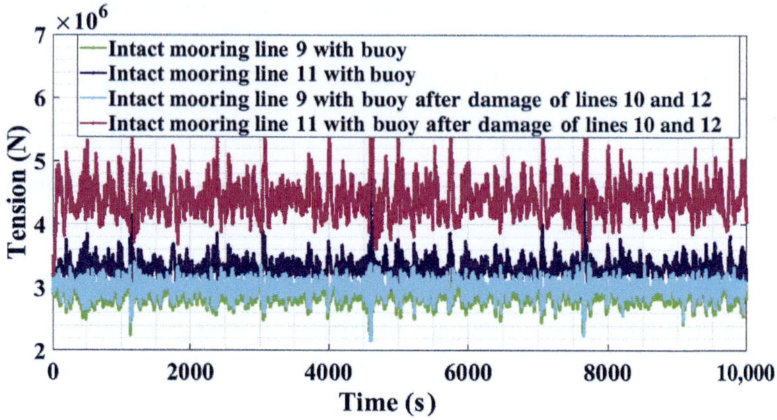

Figure 4.12 Dynamic tension variations in mooring lines (9, 11) for 0-degree wave heading.

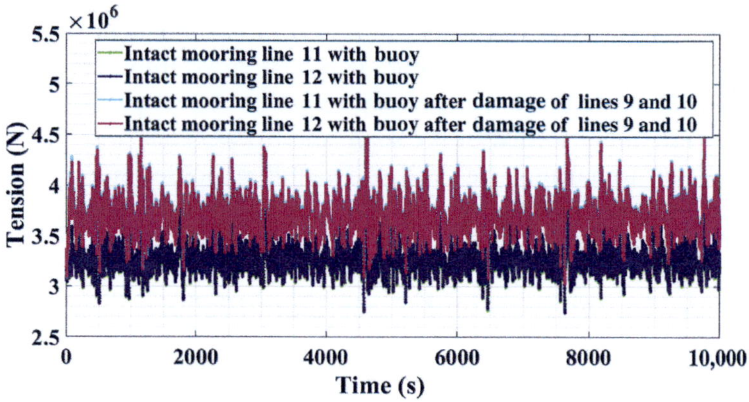

Figure 4.13 Dynamic tension variations in mooring lines (11, 12) for 45-degree wave heading.

to the failure of the adjacent mooring lines (15, 16). These variations are expected to cause a fatigue failure as they are instantaneous and cyclic. Hence, fatigue analysis is carried out to estimate the service life of the platform under such postulated failure conditions.

The fatigue life of mooring lines, both under intact and postulated failure conditions, is estimated using Palmgren–Miners rule. The dynamic tension variation in the moorings, under the motion of semisubmersible, induces a large number of stress cycles, resulting in fatigue damage. The S–N curve approach is used to estimate the number of cycles required to

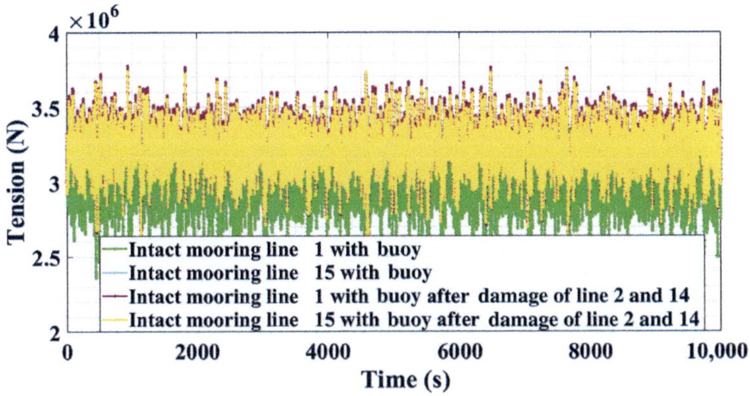

Figure 4.14 Dynamic tension variations in mooring lines (1, 15) for 90-degree wave heading.

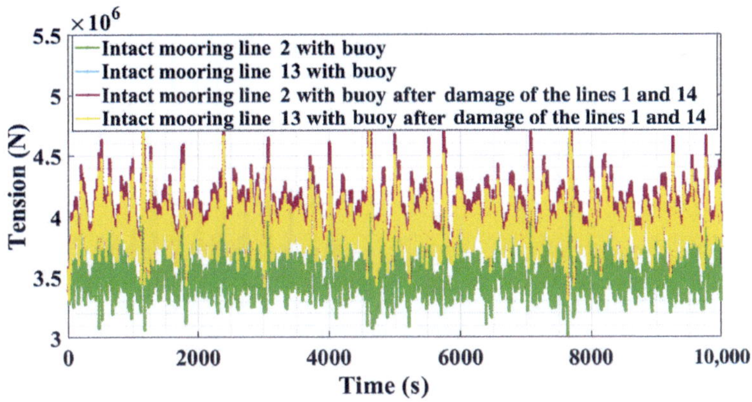

Figure 4.15 Dynamic tension variations in mooring lines (2, 13) for 0-degree wave heading.

cause fatigue damage. The pretension of the mooring causes stress response of the mooring lines to be a nonzero-mean process. After computing the stress ranges and averages of the cycle, each cycle is subsequently converted into an equivalent stress range of a zero-mean process using the Goodman diagram (Chandrasekaran, 2015b). According to the Palmgren–Miners rule, fatigue failure of the mooring line will occur when the strain energy of variable amplitude (n) is equal to that of the constant amplitude cycle (N); it is given in the following as:

$$D = \sum_{i=1}^{m} \frac{n_i}{N_i} \tag{4.10}$$

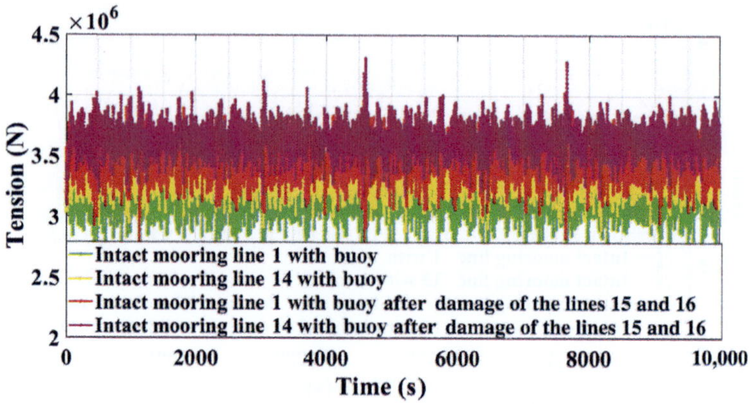

Figure 4.16 Dynamic tension variations in mooring lines (1, 14) for 0-degree wave heading.

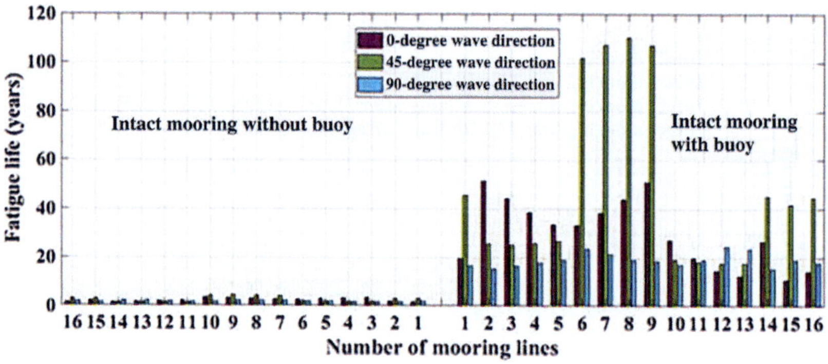

Figure 4.17 Fatigue life of intact mooring lines for a 10-year return period.

where n_i is the number of cycles per year for a given tension range interval, and N_i is the cycles to failure under tension range "i" according to the S–N curve. The parameters for the S–N curve are chosen according to the standards (Det Norske Veritas, 2005). Fatigue life is then calculated for the postulated failure condition. It is later extrapolated for 10,000 s to obtain the fatigue life of the mooring lines. Figs. 4.17 and 4.18 show the fatigue life expected for mooring lines, both under intact and failed condition, respectively, in 1500 m water depth. Both are estimated in the presence of buoy. It is seen from the figures that the inclusion of submerged buoys increased the fatigue life significantly, as they can be compared with that of the case without the buoy. It is also important to note that entire lines without buoy showed lesser fatigue life in comparison to those with a submerged buoy.

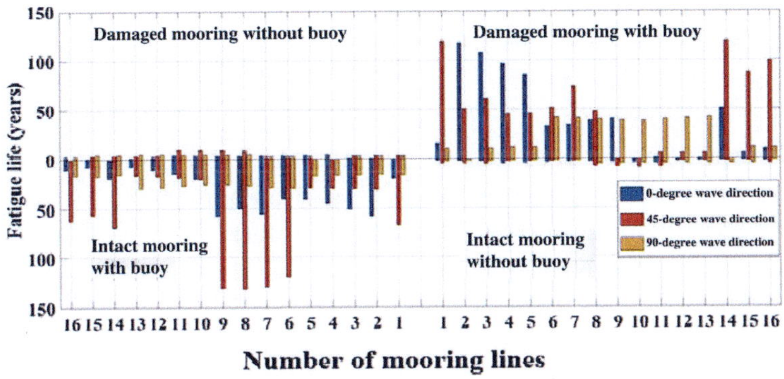

Figure 4.18 Fatigue life under postulated failure for a 10-year return period.

Coupled dynamic analysis of a semisubmersible with symmetric spread mooring, under postulated failure condition, showed that the mean response of the platform marginally increases with the inclusion of buoy. It is due to the decrease of the restoration force, but the responses are within the permissible limits. Dynamic tension variations caused in the mooring under the postulated failure condition resulted in fatigue failure. With the inclusion of the submerged buoy in the mooring system, there is a considerable decrease in tension variation in the mooring lines, increasing fatigue life. It is seen from the studies that despite the postulated failure induced in a mooring, the adjacent line remains unaffected due to a strong coupling motion of the platform.

4.9 Offshore triceratops with elliptical buoyant legs

4.9.1 Triceratops

Offshore triceratops is one of a new steel compliant structure found suitable for oil and gas exploration in ultradeepwater. It consists of three buoyant legs attached to the deck by ball joints that partially isolate the floor from the buoyant legs and constitute the novelty. Buoyant legs are anchored to the seabed by taut-moored tethers. The deck is above the mean sea level (MSL) with enough clearance, waves will interact only with the buoyant legs. As they are connected to the floor by ball joints, rotation of the buoyant legs will not be transferred to the deck; only displacements will be transferred (Chandrasekaran and Nagavinothini, 2019a). As discussed earlier, the novelty in the design of new-generation offshore platforms arises from their geometric form. While buoyant legs are generally circular in cross section, a novel form of the elliptical cross section is presented for discussion. A detailed dynamic analysis of triceratops with buoyant legs of the elliptical cross section is discussed henceforth.

A severe combination of environmental and accidental loads in a marine environment demands an adaptive structural form. Offshore triceratops is one of the new and innovative architectural types of offshore compliant structures, which is found more adaptable to ultradeepwaters (Chandrasekaran and Nagavinothini, 2018a,b; Nagavinothini and Chandrasekaran, 2019). Detailed response analyzes of triceratops validate its capacity to withstand wave, wind, current, and ice loads in ultradeepwaters (Chandrasekaran and Nagavinothini, 2018a,b, 2019a,b). With the inclusion of an elliptical section in the European Norm 10210 (CEN EN and EN, 2006; Silvestre, 2008; Chatjigeorgiou, 2011), the use of steel elliptical hollow sections is becoming popular due to their viable applications. In the case of an array of elliptical cylinders, complex hydrodynamics affects the total response of the structure significantly. They induce sway forces even under unidirectional waves, acting parallel to the longitudinal axis of the ellipse (Chatjigeorgiou and Mavrakos, 2010; Chatjigeorgiou and Katsardi, 2018). Concerning buckling behavior, elliptical shells with moderate-to-high eccentricity ($a/b \geq 2$) showed a higher ultimate load capacity in comparison to their buckling load capacity. It is because elliptical sections possess a stable, postbuckling behavior than that of circular cylindrical shells (Tennyson et al., 1971).

The conceptual model of triceratops, which is considered for the numerical analysis, is shown in Fig. 4.19. As triceratops is still in the preliminary stage, the model suitable for ultradeepwater applications is developed based on the dimensions of Perdido spar that is commissioned in the Gulf of Mexico at 2438 m water depth (Liapis et al., 2010). The geometric properties of the triceratops with cylindrical buoyant legs are listed in Table 4.11. Buoyant legs of elliptical cross section, with different eccentricities (a/b), are considered in the study. The area of the elliptical cross section is maintained as same as that of the circular ones of diameter 15.0 m. It is required to ensure equal buoyancy of the platform. Details of the elliptical cross section, discussed in the study, are listed in Table 4.12.

4.9.2 Numerical analyses

Numerical investigations are carried out of offshore triceratops with buoyant legs of elliptical cross section. The shape with different eccentricities is studied as parametric variations. Deck and the buoyant legs are mutually

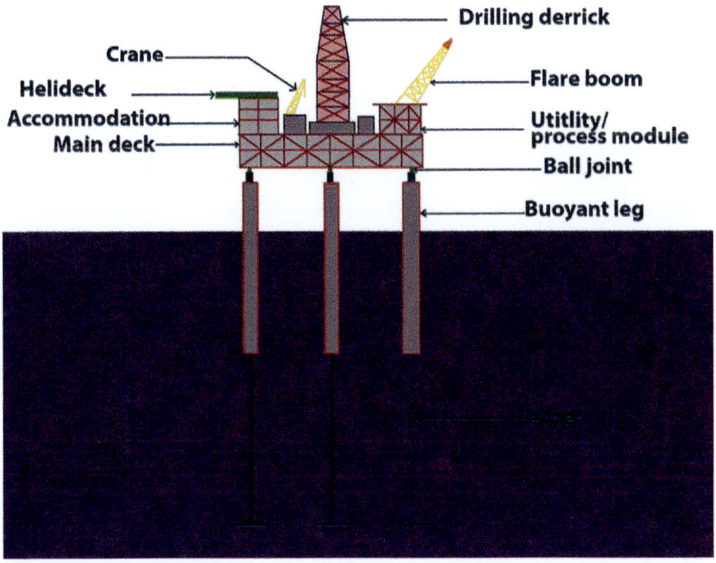

Figure 4.19 Model of triceratops.

Table 4.11 Triceratops with circular buoyant leg.

Description	Quantity
Water depth	2400 m
Unit weight of material	7850 kg/m^3
Unit weight of seawater	1025 kg/m^3
Diameter of leg	15 m
c/c Distance between the buoyant legs	61.77 m
Freeboard	20.24 m
Draft	154 m
Tether length	2246 m
Vertical center of gravity of buoyant leg	−112.74 m
Metacentric height	35.83 m
Self-weight + payload	562.42 MN
Buoyancy force	820.93 MN
Total tether force	258.49 MN
Area of deck	3933 m^3
Area of tether	2.36 m^2
Stiffness of tethers	0.22 GN/m

connected using ball joints that restrict the transfer of rotation but allow the transfer of translation between them. Subsequently, buoyant legs are anchored to the seabed using taut-moored tethers that are modeled as linear cables. Buoyant legs are meshed using quadrilateral plate elements

Table 4.12 Details of elliptical buoyant leg.

Cross section	Eccentricity (*a*/*b*)	Semimajor axis *a* (m)	Semiminor axis *b* (m)
Circular	1	7.5	7.5
Ellipse 1	1.5	9.186	6.124
Ellipse 2	2	10.567	5.263
Ellipse 3	2.5	11.859	4.743

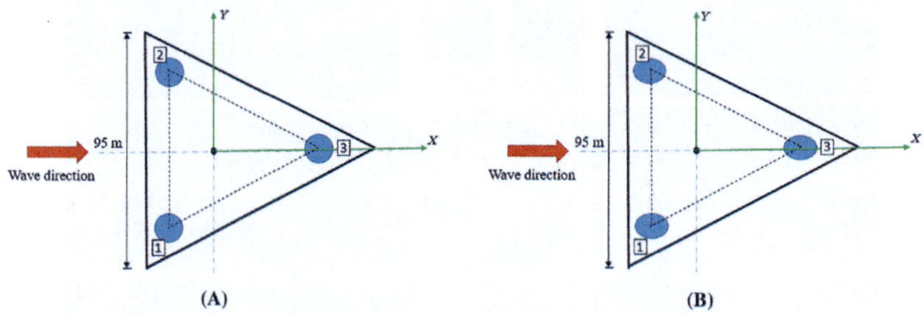

Figure 4.20 Plan view of triceratops: (A) circular and (2) ellipse.

Table 4.13 Weight of triceratops (MN) with buoyant legs of different cross sections.

Description	Circular	Ellipse 1	Ellipse 2	Ellipse 3
Weight of each buoyant leg	157.45	159.31	162.55	167.05
Total self-weight + payload	562.42	564.28	567.52	572.02
Total tether force	258.49	256.63	253.39	248.89

with four nodes to enhance the accuracy in the finite element analyses. As these elements are capable of generating both hydrostatic forces and pressure, they are found to be more appropriate. The deck is meshed using triangular and quadrilateral plate elements. Fig. 4.20 shows the plan of triceratops with buoyant legs of both cylindrical and elliptical cross sections.

Change in the cross section of the buoyant legs influences the self-weight of the triceratops. It, in turn, affects the buoyancy and the initial pretension in each tether. Table 4.13 summarizes these variations. In comparison to that of the platform with circular buoyant legs, there is an increase in the total weight. Further, there is also a significant reduction in the tether force of the platform with the buoyant legs of elliptical cross section. Reduction in the tether forces aids an extended service life of tethers. Table 4.14 summarizes the natural frequencies of triceratops in different degrees-of-freedom with buoyant legs of the different cross sections. It is seen from the table that the natural frequencies of the platform

Table 4.14 Natural frequency of tethered triceratops.

Degree-of-freedom	Circular (rad/s)	Ellipse 1 (rad/s)	Ellipse 2 (rad/s)	Ellipse 3 (rad/s)
Surge	0.029	0.029	0.028	0.028
Sway	0.029	0.029	0.028	0.028
Heave	1.461	0.829	0.829	0.914
Roll	1.013	0.980	0.991	0.942
Pitch	1.013	0.980	0.991	0.942
Yaw	0.029	0.029	0.028	0.028

with buoyant legs of elliptical cross section do not vary in the flexible degrees-of-freedom in comparison to that of the circular ones. There is a significant variation only in the stiff degrees-of-freedom such as heave, roll, and pitch. It is also important to note the fact that natural frequencies of these stiff degrees-of-freedom lie within a closer range to that of the dominant wave frequencies encountering the platform. The design is conceptualized in this desired manner to avoid higher response amplitudes by making these degrees-of-freedom stiff.

Hydrodynamic loads on offshore structures are caused by water particle kinematics, structural motion, and wave−structure interaction. Significant changes are observed in the hydrodynamic characteristics of the platform with the buoyant legs of elliptical cross section. Hydrodynamic forces on the platform consist of Froude−Krylov force and diffraction forces. While the Froude−Krylov force is the first-order wave incident force, diffraction force is the force induced by the wave disturbance due to the presence of the structure. Variations in both the diffraction and Froude−Krylov forces (and moment in case of yaw) in the buoyant leg 1 with different cross sections are shown in Figs. 4.21 and 4.22, respectively. Plots show the variations of these forces only in flexible degrees-of-freedom such as surge, sway, and yaw. The maximum forces in both the surge and sway degrees-of-freedom are observed closer to half of the natural heave frequency under all wave incident angles. It is due to a strong coupling that exists between this degrees-of-freedom. However, in the yaw degree-of-freedom, the maximum moment is observed closer to its natural frequency. It is interesting to note that no such behavior is seen in the platform with buoyant legs of circular cross section.

Variations in the diffraction and Froude−Krylov forces in the buoyant leg 1 of triceratops with the elliptical cross section show that the maximum force developed in the surge degree-of-freedom is comparatively

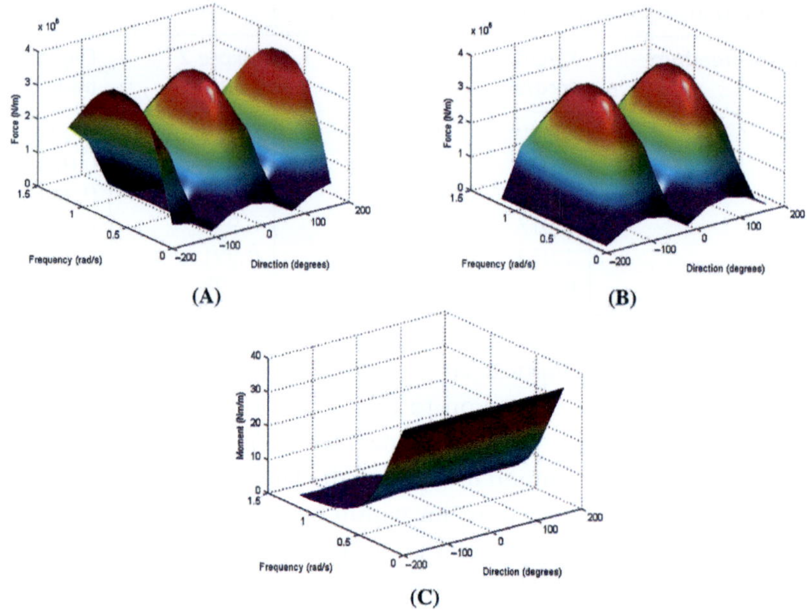

Figure 4.21 Diffraction and Froude–Krylov forces in circular buoyant legs: (A) surge, (B) sway, and (C) yaw.

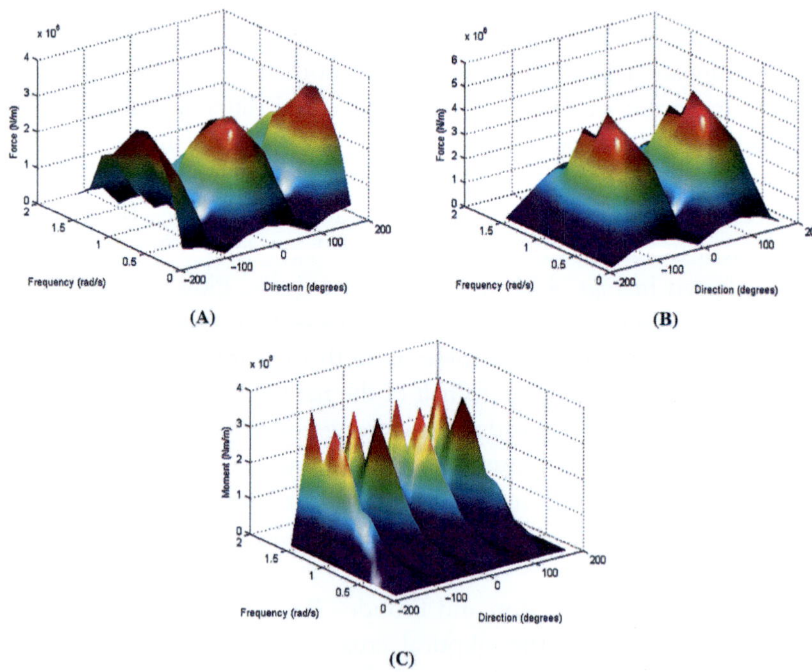

Figure 4.22 Diffraction and Froude–Krylov forces in the elliptical buoyant leg (ellipse 1): (A) surge, (B) sway, and (C) yaw.

lesser than that of the buoyant leg with a circular cross section. Force peaks in both the surge and sway plots are observed in the neighborhood of natural heave frequency. Unlike the circular buoyant leg case, several distinct peaks at various frequencies are not observed. Force in the sway degree-of-freedom increases due to the increase in the size of the leg along the *X*-direction. It encounters a higher waterplane area. With the change in the cross section of the buoyant leg, there is a shift in the maximum yaw moment. It occurs toward the higher frequencies. Minimum yaw moments occur in the directions parallel to the surge and sway degrees-of-freedom, whereas maximum moments occur at ±45 and ±135 degrees.

The dynamic response of the platform in investigated under two different sea states, as shown in Table 4.15. The total deck response in different degrees-of-freedom at zero degree incident wave angles is given in Table 4.16. It is observed that the response increases with the increase in the roughness of the sea state. Deck response decreases with the increase in the eccentricity of the shape of the elliptical buoyant legs. It is further

Table 4.15 Characteristics of sea states.

Sea state description	Douglas sea state	Wave height H (m)	Wave period T (s)	Wind speed (m/s)
Rough	4	2.0	5.0	10
High	6	4.0	5.0	20

Table 4.16 Deck response under different sea states.

Sea state	Degree-of-freedom	Circular	Ellipse 1	Ellipse 2	Ellipse 3
Rough	Surge (m)	0.4047	0.5098	0.4296	0.4067
	Sway (m)	0.0192	0.1434	0.0348	0.1537
	Heave (m)	0.0012	0.0012	0.0011	0.0011
	Roll (degrees)	0.0035	0.0031	0.0027	0.0028
	Pitch (degrees)	0.0037	0.0038	0.0036	0.0034
	Yaw (degrees)	0.0253	0.3688	0.0815	0.0293
High	Surge (m)	0.8471	1.6684	0.8189	0.9197
	Sway (m)	0.0778	1.0732	0.0514	0.5438
	Heave (m)	0.0015	0.0014	0.0014	0.0013
	Roll (degrees)	0.0024	0.0038	0.0021	0.0035
	Pitch (degrees)	0.0055	0.0048	0.0051	0.0044
	Yaw (degrees)	0.0711	1.9830	0.1178	0.9904

seen that the total surge response of the deck under ellipse 1 increases by about 26% in comparison to that of the platform with circular buoyant legs. Under circular buoyant legs, sway response of the floor is about 4.70% of that of the surge while sway response is lesser when compared to the surge response only in case of ellipse 2; transverse vibrations are also lower in this case. Transverse vibration is found to be higher in the case of ellipse 3, where the sway response is about 38% of that of the surge. Heave, roll, pitch, and yaw responses of the deck decrease with the increase in the eccentricity of the shape of elliptical buoyant legs. In high sea state, surge response of the deck in case of ellipse 2 is marginally lesser than that of the circular ones, whereas it is higher for the other two cases. Transverse vibration is also comparatively lower in the case of ellipse 2.

Further, sway response is only about 6.3% of that of the surge in the case of ellipse 2, whereas it is about 9% in the platform with circular buoyant legs. Reduced heave, roll, and pitch responses are also observed in the case of ellipse 2 in comparison to the circular ones. However, there is an increase in the yaw response by about 66%. It is therefore clear that the shape of the buoyant legs does not influence the stiff degrees-of-freedom; only flexible degrees-of-freedom such as surge, sway, and yaw show more considerable variations with the change in the eccentricity of the buoyant legs.

Under the action of wave loads and current, the tension in the tethers varies continuously. It is because surge/sway responses cause set down in the heave degree-of-freedom, resulting in the dynamic variation in buoyancy. It affects the axial tension in tethers, producing a dynamic tension variation. The motion of triceratops under wave action alters the initial tension in the tethers used for position restraining the buoyant legs; these variations under different cases under both rough and high sea states are listed in Table 4.17.

Table 4.17 Tether tension variations.

Sea state	Tether tension	Circular	Ellipse 1	Ellipse 2	Ellipse 3
Rough	Maximum (MN)	29.04	28.83	28.42	27.97
	Minimum (MN)	28.40	28.19	27.83	27.38
	Variation (%)	2.20	2.20	2.10	2.10
High	Maximum (MN)	29.15	28.89	28.48	28.02
	Minimum (MN)	28.29	28.08	27.84	27.32
	Variation (%)	3.00	2.80	2.30	2.50

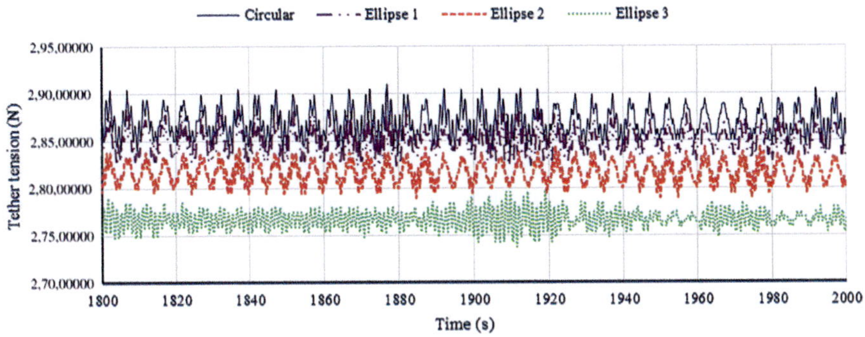

Figure 4.23 Tether tension variation in buoyant leg 1 under high sea state.

It is seen from the table that the dynamic tension variation increases with the increase in the roughness of the sea state. In the rough sea state, there is no significant difference in the tension variation to the changes in the cross section of the buoyant legs. However, under the high sea state, a maximum tension variation is observed in the circular legs. In contrast, a minimum difference is seen in the elliptical legs with eccentricity 2.0, which is an added advantage to the cross section of the ellipse 2 configuration. Tether tension response of different buoyant leg cases under high sea state is shown in Fig. 4.23.

Free-oscillation studies conducted on tethered triceratops with buoyant legs of different cross sections showed that the change in the shape of the cross section influences the natural frequency in stiff degrees-of-freedom. The diffraction and Froude–Krylov forces acting on the buoyant legs increase with an increase in the eccentricity of the shape of elliptical sections. Further, it is seen that the elliptical buoyant leg with eccentricity 2.0 is comparatively advantageous than that of other cross sections. It showed a reduced deck response with increased stability in different degrees-of-freedom. Also, a reduced tether tension variation in the case of ellipse 2 adds to its advantage.

Exercise

Part A: Objective-type questions

1. Ocean structures are more focused on being __________ (form-dominant) rather than strength-based design.
2. List a few materials used for the construction of ocean structures.

   A wide range of materials such as composites, concrete, mild steel, copper, aluminum, fiber-reinforced plastics, titanium, and other buoyancy materials are preferred. The choice of their use depends on the application requirements as all the above cannot be used for the same purpose.

3. List a few parameters that govern the choice of materials for ocean structures.

Under the given choice of a wide variety of materials available for use in offshore and marine structures, a few parameters govern their choice: type of applications, where a specific material is chosen, structural properties under regular, cyclic loads, high and very low temperature, recycling characteristics, nontoxic nature and biofriendly characteristics, sustainability for the service life of the structure under marine environment, and the code compliance. It is, therefore, necessary to ensure that the chosen material for construction should possess a large availability in the desired shape, size, and modular sections, apart from being cost-effective.

4. __________ (form-dominant) capability of offshore compliant structures enables an effective load disbursement.

5. ____________ (mechanical properties) are generally considered as essential indices to study the behavior of metals under loads.

6. List a few essential mechanical properties.

Some of the essential properties are strength, hardness, elasticity, plasticity, brittleness, ductility, and malleability.

7. _________________ (dynamic tension variations) of top-tensioned risers, held down from a semisubmersible, can result in fatigue failure.

8. _________________ (ductile behavior) of the steel is well understood from the stress–strain curve plotted under the uniaxial tensile load.

9. Plot the variation of carbon steel under high temperature.

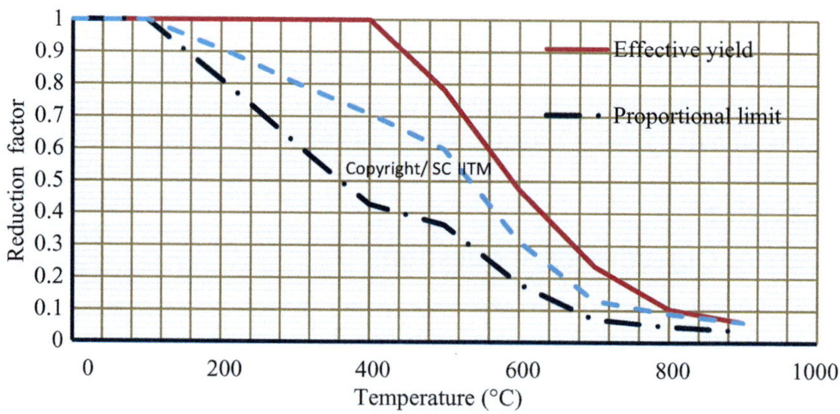

10. Which is the specific property of steel at high temperatures that is discouraging?

Although steel is one of the most preferred construction materials for marine risers, strength degradation and decrystallization of steel at higher temperatures remain a significant challenge.

11. _______________ (delamination) of composites limits their use as an engineering material.

12. What issues a composite material face under high temperature?

Under high temperature and pressure, thermal stresses occur at the bonding interface due to the differences in their coefficients of thermal expansion. It leads to a crack formation and results in delamination at the interface.

13. Name two characteristics of steel that are modified while manufacturing FGM.

Specific properties such as the yield strength and tangent modulus of the individual materials are technically manipulated to arrive at a functionally graded combination.

14. Draw a cross section of FGM riser.

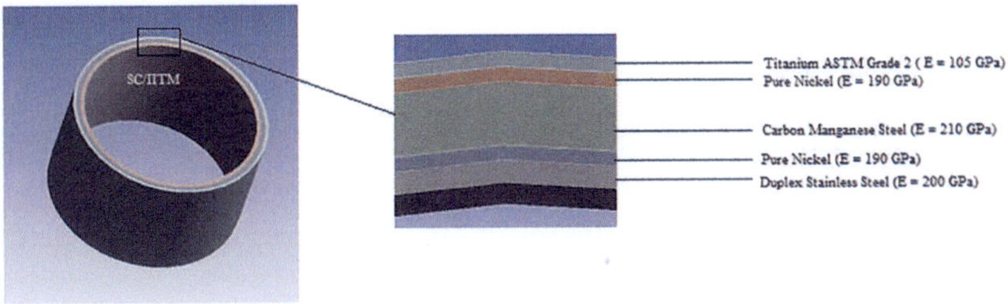

15. Where are titanium mainly used?

The majority of offshore applications use titanium for submarine ball valves, pumps, and heat exchangers.

16. In the case of hot working with titanium, what are the specific problems faced?

In the case of hot working with titanium, absorption of certain gases such as hydrogen (above 148°C), oxygen (above 700°C), and nitrogen (above 815°C) causes embrittlement. Therefore to avoid hydrogen absorption, hot working is carried out in an oxidizing atmosphere while the oxygen-embrittled layer can be removed after hot working.

17. Exposure to _____________ (ultraviolet rays) causes brittleness to fiberglass composites.

18. ____________ (buoyancy materials) have specific gravity considerably lesser than water.
19. There are ____________ (six) degrees-of-freedom for a floating body in space.
20. ____________ (buoys), if provided at various sections along the steel moorings, found to increase their fatigue life.

Part B: Descriptive-type questions

1. Write a brief note on the selection of materials for ocean structures.
2. How does the marine environment affect the structural characteristics of steel?
3. Compare the properties of individual materials used to fabricate FGM.
4. Write a brief note on titanium.
5. How are composites useful for the construction of structures in the marine environment?
6. Write a short note on nonferrous materials used in ocean structures.
7. Write a summary of syntactic foams.
8. Explain how members with elliptical cross section are useful in alleviating wave loads acting on them.
9. Discuss the properties of carbon steel under high temperature.
10. What factors influence the selection of materials in ocean structures?

References

Abbassian, F., Moros, T., 1996. The use of air-bubble spoilers to suppress vortex-induced vibrations of risers. SPE Prod. Facil. 11 (01), 35–40. Available from: https://doi.org/10.2118/25036-PA.

Albino, J.C.R., Almeida, C.A., Menezes, I.F.M., Paulino, G.H., 2018. Co-rotational 3D beam element for nonlinear dynamic analysis of risers manufactured with functionally graded materials (FGM). Eng. Struct. 173, 283–299. Available from: https://doi.org/10.1016/j.engstruct.2018.05.092.

Anagnostopoulos, P., Bearman, P.W., 1992. Response characteristics of a vortex-excited cylinder at low Reynolds numbers. J. Fluids Struct. 6 (1), 39–50.

Anil, B.J., 2014. Vortex Vibrations of Inclined Cylinder in Flow (Master's thesis). University of Massachusetts. Amherst, MA.

Aslan, Z., Daricik, F., 2016. Effects of multiple delaminations on the compressive, tensile, flexural, and buckling behavior of E-glass/epoxy composites. Composites, B: Eng. 100, 186–196. Available from: https://doi.org/10.1016/j.compositesb.2016.06.069.

Baarholm, G.S., Larsen, C.M., Lie, H., 2006. On fatigue damage accumulation from in-line and cross-flow vortex-induced vibrations on risers. J. Fluids Struct. 22 (1), 109–127. Available from: https://doi.org/10.1016/j.jfluidstructs.2005.07.013.

Bai, Y., Bai, Q., 2005. Subsea Pipelines and Risers. Elsevier, ISBN: 0-080-4456-67, ISBN.

Bearman, P.W., 1984. Vortex shedding from oscillating bluff bodies. Annu. Rev. Fluid Mech. 16 (1), 195–222.

Bermingham, M.J., Kent, D., Zhan, H., StJohn, D.H., Dargusch, M.S., 2015. Controlling the microstructure and properties of wire arc additive manufactured Ti–6Al–4V with trace boron additions. Acta Mater. 91, 289–303. Available from: https://doi.org/10.1016/j.actamat.2015.03.035.

Blevins, D., 1994. Flow Induced Vibration, second ed. Krieger Publishing, Malabar, FL.

Bloor, M.S., 1964. The transition to turbulence in the wake of a circular cylinder. J. Fluid Mech. 19, 290–304.

Brika, D., Laneville, A., 1993. Vortex-induced vibrations of a long flexible circular cylinder. J. Fluid Mech. 250, 481–508.

CEN EN, 2006. Hot Finished Hollow Structural Sections of Non-alloy and Fine Grain Steels—Part 2: Tolerances Dimensions and Sectional Properties. Comité Européen de Normalisation Brussels, Belgium.

Chakrabarti, S.K., 2002. The Theory and Practice of Hydrodynamics and Vibration. World Scientific Publishing Company, Singapore.

Chakrabarti, S., 2005. Handbook of Offshore Engineering, 2. Elsevier, UK, ISBN: 9780080443812, ISBN.

Chakrabarti, S.K., Frampton, R.E., 1982. Review of riser analysis techniques. Appl. Ocean Res. 4 (2), 73–90. Available from: https://doi.org/10.1115/1.3604697.

Chandrasekaran, S., 2014. Advanced Theory on Offshore Plant FEED Engineering. Changwon National University Press, Republic of South Korea, ISBN: 978-899-69-7928-9, p. 237, ISBN.

Chandrasekaran, S., 2015a. Advanced Marine Structures. CRC Press, Boca Raton, FL, ISBN: 9781498739689, ISBN.

Chandrasekaran, S., 2015b. Dynamic Analysis and Design of Ocean Structures. Springer, India, ISBN: 978-81-322-2276-7, ISBN.

Chandrasekaran, S., 2016a. Health, Safety and Environmental Management for Offshore and Petroleum Engineers. John Wiley and Sons, UK, ISBN: 978-11-192-2184-5, ISBN.

Chandrasekaran, S., 2016b. Offshore Structural Engineering: Reliability and Risk Assessment. CRC Press, Boca Raton, FL, ISBN: 978-14-987-6519-0, ISBN.

Chandrasekaran, S., 2017. Dynamic Analysis and Design of Ocean Structures, second ed. Springer, Singapore.

Chandrasekaran, S., 2019a. Advanced Steel Design of Structures. CRC Press, Boca Raton, FL, ISBN: 978-036-72-3290-0, ISBN.

Chandrasekaran, S., 2019b. Structural Health Monitoring With Application to Offshore Structures. World Scientific Publishing Co., Singapore, ISBN: 978-981-12-0108-0, ISBN.

Chandrasekaran, S., Bhattacharyya, S.K., 2012. Analysis and Design of Offshore Structures With Illustrated Examples. Human Resource Development Center for Offshore and Plant Engineering (HOPE Center), Changwon National University Press, Republic of Korea, ISBN: 978-899-63-9155-5, p. 285, ISBN.

Chandrasekaran, S., Gaurav, S., 2017. Design Aids for Offshore Structures Under Special Environmental Loads Including Fire Resistance. Springer, Singapore, ISBN: 978-981-10-7608-4, ISBN.

Chandrasekaran, S., Jain, A.K., 2016. Ocean Structures: Construction, Materials and Operations. CRC Press, Boca Raton, FL, ISBN: 978-14-987-9742-9, ISBN.

Chandrasekaran, S., Madhavi, N., 2014a. Numerical study on geometrical configurations of perforated cylindrical structures. J. Perform. Constr. Faci. 30 (1), 04014185. Available from: https://doi.org/10.1061/(ASCE)CF.1943-5509.0000687.

Chandrasekaran, S., Madhavi, N., 2014b. Retrofitting of offshore structural member using perforated cylinders. SFA Newsl. 13, 10–11.

Chandrasekaran, S., Madhavi, N., 2015a. Design aids for offshore structures with perforated members. J. Ship Offshore Struct. 10 (2), 183–203. Available from: https://doi.org/10.1080/17445302.2014.918309.

Chandrasekaran, S., Madhavi, N., 2015b. Retrofitting of offshore cylindrical structures with different geometrical configuration of perforated outer cover. Int. J. Shipbuild. Prog. 62 (1–2), 43–56. Available from: https://doi.org/10.3233/ISP-150115.

Chandrasekaran, S., Madhavi, N., 2015c. Flow field around a outer perforated circular cylinder under regular waves: numerical study. J. Mar. Syst. Ocean Technol. 10 (2), 91–100. Available from: https://doi.org/10.1007/s40868-015-0008-1.

Chandrasekaran, S., Mayanak, S., 2017. Dynamic analyses of stiffened triceratops under regular waves: experimental investigations. Ships Offshore Struct. 12 (5), 697–705. Available from: https://doi.org/10.1080/17445302.2016.1200957.

Chandrasekaran, S., Merin, T., 2016. Suppression system for offshore cylinders under vortex induced vibration. Vibroeng. Procedia 7, 01–06.

Chandrasekaran, S., Nagavinothini, R., 2018a. Dynamic analyses and preliminary design of offshore triceratops in ultra-deep waters. Innovative Infrastruct. Solut. 3 (1), 16. Available from: https://doi.org/10.1007/s41062-017-0124-1.

Chandrasekaran, S., Nagavinothini, R., 2018b. Tether analyses of offshore triceratops under the wind, wave, and current. Mar. Syst. Ocean Technol. 13 (1), 34–42. Available from: https://doi.org/10.1007/s40868-018-0043-9.

Chandrasekaran, S., Nagavinothini, R., 2019a. *Compliant Platforms: Analysis, Design and Experimental Studies*. Wiley, UK, ISBN: 9781119669777, ISBN.

Chandrasekaran, S., Nagavinothini, R., 2019b. Tether analyses of offshore triceratops under ice force due to continuous crushing. Innovative Infrastruct. Solut. 4 (1), 25. Available from: https://doi.org/10.1007/s41062-019-0212-5.

Chandrasekaran, S., Ranjani, R., Kumar, D., 2015a. Response control of tension leg platform with passive damper: experimental investigations. J. Ships Offshore Struct. 12 (2), 171–181. Available from: https://doi.org/10.1080/17445302.2015.1119666.

Chandrasekaran, S., Natarajan, M., Sreeramulu, L.R., 2015b. Estimation of force reduction on ocean structures with perforated members. In: Proceedings of the 34th International Conference on Ocean, Offshore and Arctic Engineering. OMAE2015-41153, May 31–June 5, 2015, American Society of Mechanical Engineers, St. John's, NL, pp. V007T06A035–V007T06A035.

Chandrasekaran, S., Hari, S., Murugaiyan, M., 2019. Wire-arc additive manufacturing of functionally-graded material for marine riser applications. In: Proceedings of the I-OCEANS 2019. 5–7 Aug 2019, Universiti Teknologi Malaysia, Terengganu, Malaysia.

Chatjigeorgiou, I.K., 2011. Three-dimensional wave scattering by arrays of elliptical and circular cylinders. Ocean Eng. 38 (13), 1480–1494. Available from: https://doi.org/10.1016/j.oceaneng.2011.07.001.

Chatjigeorgiou, I.K., Katsardi, V., 2018. Hydrodynamics and near trapping effects in arrays of multiple elliptical cylinders in waves. Ocean Eng. 157, 121–139. Available from: https://doi.org/10.1016/j.oceaneng.2018.03.045.

Chatjigeorgiou, I.K., Mavrakos, S.A., 2010. An analytical approach for the solution of the hydrodynamic diffraction by arrays of elliptical cylinders. App. Ocean Res. 32 (2), 242–251. Available from: https://doi.org/10.1016/j.apor.2009.11.004.

Chen, P., Chai, S., Ma, J., 2011. Performance evaluations of taut-wire mooring systems for the deep-water semi-submersible platform. In: Proceedings of the 30th International Conference on Ocean, Offshore and Arctic Engineering. pp. 207–215. https://doi.org/10.1115/OMAE2011-49281.

Chen, X., Li, J., Cheng, X., He, B., Wang, H., Huang, Z., 2017. Microstructure and mechanical properties of the austenitic stainless steel 316L fabricated by gas metal arc additive manufacturing. Mat. Sci. Eng.: A 703, 567–577. Available from: https://doi.org/10.1016/j.msea.2017.05.024.

Chia-Rong, C., Hamn-Ching, C.C., 2016. Simulation of vortex-induced motions of a deep draft semi-submersible in current. J. Ocean Eng. 118, 107–116.

DebRoy, T., Wei, H.L., Zuback, J.S., Mukherjee, T., Elmer, J.W., Milewski, J.O., et al., 2018. Additive manufacturing of metallic components–process, structure, and properties. Prog. Mater. Sci. 92, 112–224. Available from: https://doi.org/10.1016/j.pmatsci.2017.10.001.

Det Norske Veritas, 2002. DNV-Recommended Practice 201: Design of Titanium Risers. DNV GL, Oslo, pp. 1–26.

Det Norske Veritas, 2005. DNV-Recommended Practice C203: Fatigue Design of Offshore Steel Structures. DNV GL, Oslo, pp. 14–16.

Det Norske Veritas, 2012. DNV-OS-F101 Submarine Pipeline Systems. DNV GL, Oslo, pp. 1–367.

Det Norske Veritas, 2016. DNV-Recommended Practice F205: Global Performance Analysis of Deepwater Floating Structures. DNV GL, Oslo.

Det Norske Veritas, 2018. DNVGL-RP-F112 Duplex Stainless Steel – Design Against Hydrogen-Induced Stress Cracking. DNV GL, Oslo, pp. 1–44.

Elosta, H., Huang, S., Incecik, A., 2014. Wave loading fatigue reliability and uncertainty analyses for geotechnical pipeline models. Ships Offshore Struct. 9 (4), 450–463. Available from: https://doi.org/10.1080/17445302.2013.834168.

Evan, J.L., 2007. Airfoil Vortex Induced Vibration Suppression Devices (Master of Science thesis). Massachusetts Institute of Technology, Dept. of Mechanical Engineering.

Feng, C.C., 1968. The Measurement of Vortex Induced Effects in Flow Past Stationary and Oscillating Circular and D-Section Cylinders (Master of Science thesis). University of British Columbia.

Finn, L.D., Gupta, H., Maher, J.V., 2003. The cell spar and vortex induced vibrations. In: Proceedings of Offshore Technology Conference. OTC15244, Houston, TX.

Gabbai, R.D., Benaroya, H., 2005. An overview of modeling and experiments of vortex-induced vibration of circular cylinders. J. Sound Vib. 282, 575–616.

Gao, J.W., Wang, C.Y., 2000. Modeling the solidification of functionally graded materials by centrifugal casting. Mater. Sci. Eng.: A 292 (2), 207–215. Available from: https://doi.org/10.1016/S0921-5093(00)01014-5.

Gao, Y., Zong, Z., Sun, L., 2011. Numerical prediction of fatigue damage in steel catenary riser due to vortex-induced vibration. J. Hydrodyn. 23 (2), 154–163. Available from: https://doi.org/10.1016/S1001-6058(10)60099-6.

Gao, Y., Fu, S., Ren, T., Xiong, Y., Song, L., 2015. VIV response of a long flexible riser fitted with strakes in uniform and linearly sheared currents. Appl. Ocean Res. 52, 102–114. Available from: https://doi.org/10.1080/17445302.2016.1187832.

Garza-Rios, L.O., Bernitsas, M.M., Nishimoto, K., Matsuura, J., 2000. Dynamics of spread mooring systems with hybrid mooring lines. J. Offshore Mech. Arct. Eng. 122 (4), 274–281. Available from: https://doi.org/10.1115/1.1315591.

Gerrard, J.H., 1966. The mechanics of the formation region of vortices behind bluff bodies. J. Fluid Mech. 25, 401–413.

Gouveia, J., Sriskandarajah, T., Karunakaran, D., Manso, D., Chiodo, M., Zhou, D., et al., 2015. Steel catenary risers (SCRs): from design to installation of the first reeled CRA lined pipes. Part I—Risers design. In: Offshore Technology Conference. doi:10.4043/25839-MS.

Govardhan, R., Williamson, C.H.K., 2000. Modes of vortex formation and frequency response of a freely vibrating cylinder. J. Fluid Mech. 420, 85–130.

Govardhan, R., Williamson, C.H.K., 2002. Resonance forever: existence of a critical mass and an infinite regime of resonance in vortex-induced vibration. J. Fluid Mech. 473, 147–166.

Griffin, O.M., Ramberg, S.E., 1982. Some recent studies of vortex shedding with application to marine tubulars and risers. J. Energy Resour. Technol. 104, 2–13.

Griffin, O.M., Skop, R.A., Koopman, G.H., 1973. The vortex-excited resonant vibrations of circular cylinders. J. Sound. Vib. 31, 235–249.

Hallam, M.G., Heaf, N.J., Wootton, L.R., 1977. Dynamics of marine structures. CIRIA Underwater Engineering Group, Report UR8 Atkins Research and Development. London.

Hang, S.C., Hyun, S.S., Park, I.K., Jun, B.R., 2003. An experimental study for mooring effects on the stability of the spar platforms. In: Proceedings of the 13th International Offshore and Polar Engineering Conference, vol. 1. Honolulu, Hawaii, pp. 285–288.

Hassan, A., Downie, M.J., Incecik, A., Baarholm, R., Berthelsen, P.A., Pakozdi, C., et al., 2009. Contribution of the mooring system to the low-frequency motions of a semi-submersible in combined wave and current. In: ASME 2009 28th International Conference on Ocean, Offshore and Arctic Engineering. American Society of Mechanical Engineers Digital Collection, pp. 55–62. https://doi.org/10.1115/OMAE2009-79074.

Hong, K.S., Bentsman, J., 1994a. Application of averaging method for integro-differential equations to model reference adaptive control of parabolic systems. Automatica 30 (9), 1415–1419. Available from: https://doi.org/10.1016/0005-1098(94)90006-X.

Hong, K.S., Bentsman, J., 1994b. Direct adaptive control of parabolic systems: algorithm synthesis and convergence and stability analysis. IEEE Trans. Autom. Control. 39 (10), 2018–2033. Available from: https://doi.org/10.1109/9.328823.

Hong, K.S., Shah, U.H., 2018. Vortex-induced vibrations and control of marine risers: a review. Ocean Eng. 152, 300–315. Available from: https://doi.org/10.1016/j.oceaneng.2018.01.086.

Hover, F.S., Techet, A.H., Triantafyllou, M.S., 1998. Forces on oscillating uniform and tapered cylinders in crossflow. J. Fluid Mech. 363, 97–114.

Hover, F.S., Tvedt, H., Triantafyllou, M.S., 2001. Vortex induced vibrations of a cylinder with tripping wires. J. Fluid Mech. 448, 175–195.

How, B.V.E., Ge, S.S., Choo, Y.S., 2009. Active control of flexible marine risers. J. Sound Vib. 320 (4–5), 758–776. Available from: https://doi.org/10.1016/j.jsv.2008.09.011.

Huang, T., Dareing, D.W., 1968. Buckling and lateral vibration of drill pipe. J. Eng. Ind. 90 (4), 613–619. Available from: https://doi.org/10.1115/1.3604697.

Huang, W., Liu, H.X., Shan, G.M., Hu, C., 2011. Fatigue analysis of the taut wire mooring system applied for deep waters. J. China Ocean Eng. 25 (3), 413–426. Available from: https://doi.org/10.1007/s13344-011-0034-5.

Hussain, A., Nah, E., Fu, R., Gupta, A., 2009. Motion comparison between a conventional deep draft semi-submersible and a dry tree semi-submersible. In: ASME 2009 28th International Conference on Ocean, Offshore and Arctic Engineering. American Society of Mechanical Engineers Digital Collection, pp. 785–792. https://doi.org/10.1115/OMAE2009-80006.

Igarashi, T., 1986. Effect of tripping wires on the flow around a circular cylinder normal to an airstream. Bull. Jpn. Soc. Mech. Eng. 29, 2917–2924.

Jang, B.S., Kim, J.D., Park, T.Y., Jeon, S.B., 2019. FEA based optimization of semi-submersible floater considering buckling and yield strength. Int. J. Nav. Archit. Ocean Eng. 11 (1), 82–96. Available from: https://doi.org/10.1016/j.ijnaoe.2018.02.010.

Jeans, G., Grant, C., Feld, G., 2002. Improved current profile criteria for deepwater riser design. In: ASME 2002 21st International Conference on Offshore Mechanics and Arctic Engineering. American Society of Mechanical Engineers, pp. 307–311. doi:10.1115/OMAE2002-28153.

Jensen, G.A., Säfström, N., Nguyen, T.D., Fossen, T.I., 2010. A nonlinear PDE formulation for offshore vessel pipeline installation. Ocean Eng. 37 (4), 365–377. Available from: https://doi.org/10.1016/j.oceaneng.2009.12.009.

Jin, G., Takeuchi, M., Honda, S., Nishikawa, T., Awaji, H., 2005. Properties of multilayered mullite/Mo functionally graded materials fabricated by powder metallurgy processing. Mater. Chem. Phys. 89 (2–3), 238–243. Available from: https://doi.org/10.1016/j.matchemphys.2004.03.031.

Kan, W.H., Albino, C., Dias-da-Costa, D., Dolman, K., Lucey, T., Tang, X., et al., 2018. Microstructure characterization and mechanical properties of a functionally-graded NBC/high chromium white cast iron composite. Mater. Charact. 136, 196–205. Available from: https://doi.org/10.1016/j.matchar.2017.12.020.

Kawasaki, A., Watanabe, R., 1987. Finite element analysis of thermal stress of the metal/ceramic multi-layer composites with controlled compositional gradients. Jpn. Inst. Metals, J. 51, 525–529. Available from: https://doi.org/10.2320/jinstmet1952.51.6_525.

Khalak, A., Williamson, C.H.K., 1991. Motions, forces and mode transitions in vortex induced vibrations at low mass-damping. J. Fluids Struct. 13, 813–851.

Kim, C.S., Hong, K.S., 2009. Boundary control of container cranes from the perspective of controlling an axially moving string system. Int. J. Control, Autom. Syst. 7 (3), 437–445. Available from: https://doi.org/10.1007/s12555-009-0313-6.

Kim, Y., Kim, K.H., Kim, J.H., Kim, T., Seo, M.G., Kim, Y., 2011. Time-domain analysis of nonlinear motion responses and structural loads on ships and offshore structures:

development of WISH programs. Int. J. Nav. Archit. Ocean Eng. 3 (1), 37–52. Available from: https://doi.org/10.2478/IJNAOE-2013-0044.

King, R., 1977. A review of vortex shedding research and its application. J. Ocean Eng. 4, 141–171.

Koizumi, M.F.G.M., 1997. FGM activities in Japan. Composites, B: Eng. 28 (1–2), 1–4. Available from: https://doi.org/10.1016/S1359-8368(96)00016-9.

Kurian, V.J., Yassir, M.A., Harahap, I.S., 2010. Nonlinear coupled dynamic response of a semi-submersible platform. In: the 20th International Offshore and Polar Engineering Conference. International Society of Offshore and Polar engineers.

Lee, Y., Incecik, A., Chan, H.S., 2005. Prediction of global loads and structural response analysis on a multi-purpose semi-submersible. In: ASME 2005 24th International Conference on Offshore Mechanics and Arctic Engineering. American Society of Mechanical Engineers Digital Collection, pp. 3–13. https://doi.org/10.1115/OMAE2005-67003.

Leira, B.J., Meling, T.S., Larsen, C.M., Berntsen, V., Stahl, B., Trim, A., 2005. Assessment of fatigue safety factors for deep-water risers in relation to VIV. J. Offshore Mech. Arct. Eng. 127 (4), 353–358. Available from: https://doi.org/10.1115/1.2087548.

Lesage, F., Garthshore, L.S., 1987. A method of reducing drag and fluctuating side force on bluff bodies. J. Wind Eng. Ind. Aerodyn. 25, 229–245.

Liapis, S., Bhat, S., Caracostis, C., Webb, C., Lohr, C., 2010. Global performance of the Perdido spar in waves, wind and current: numerical predictions and comparison with experiments. In: ASME 2010 29th International Conference on Ocean, Offshore and Arctic Engineering. American Society of Mechanical Engineers Digital Collection, pp. 863–873. https://doi.org/10.1115/OMAE2010-21116.

Lie, H., Kaasen, K.E., 2006. Modal analysis of measurements from a large-scale VIV model test of a riser in linearly sheared flow. J. Fluids Struct. 22 (4), 557–575. Available from: https://doi.org/10.1016/j.jfluidstructs.2006.01.002.

Liu, X., Chen, G., Fu, J., Ji, J., Song, Q., Chang, Y., 2016. Analysis of the operation fatigue of deepwater drilling riser system. Open. Pet. Eng. J. 9 (1). Available from: https://doi.org/10.2174/1874834101609010279.

Lyapin, A.A., Chebakov, M.I., Dumitrescu, A., Zecheru, G., 2015. Finite-element modelling of a damaged pipeline repaired using the wrap of a composite material. Mech. Compos. Mater. 51 (3), 333–340. Available from: https://doi.org/10.1007/s11029-015-9504-9.

Ma, Y., Cuiuri, D., Hoye, N., Li, H., Pan, Z., 2015. The effect of location on the microstructure and mechanical properties of titanium aluminides produced by additive layer manufacturing using in-situ alloying and gas tungsten arc welding. Mater. Sci. Eng.: A 631, 230–240. Available from: https://doi.org/10.1016/j.msea.2015.02.051.

Madec, C., Le Gallet, S., Salesse, B., Geoffroy, N., Bernard, F., 2018. Alumina-titanium functionally graded composites produced by spark plasma sintering. J. Mater. Process. Technol. 254, 277–282. Available from: https://doi.org/10.1016/j.jmatprotec.2017.11.004.

Martina, F., Mehnen, J., Williams, S.W., Colegrove, P., Wang, F., 2012. Investigation of the benefits of plasma deposition for the additive layer manufacture of Ti–6Al–4V. J. Mater. Process. Technol. 212 (6), 13. Available from: https://doi.org/10.1016/j.jmatprotec.2012.02.002.

Mavrakos, S.A., Chatjigeorgiou, J., 1997. Dynamic behavior of deepwater mooring lines with submerged buoys. Comput. Struct. 64 (1–4), 819–835. Available from: https://doi.org/10.1016/S0045-7949(96)00169-1.

Mavrakos, S.A., Papazoglou, V.J., Triantafyllou, M.S., Hatjigeorgiou, J., 1996. Deepwater mooring dynamics. Mar. Struct. 9 (2), 181–209. Available from: https://doi.org/10.1016/0951-8339(94)00019-O.

Mazzilli, C.E., Sanches, C.T., 2011. Active control of vortex-induced vibrations in offshore catenary risers: a nonlinear normal mode approach. J. Mech. Mater. Struct. 6 (7), 1079–1088. Available from: https://doi.org/10.2140/jomms.2011.6.1079.

Modarres-Sadeghi, Y., Chasparis, F., Triantafyllou, M.S., Tognarelli, M., Beynet, P., 2011. Chaotic response is a generic feature of vortex-induced vibrations of flexible risers. J. Sound Vib. 330 (11), 2565–2579. Available from: https://doi.org/10.1016/j.jsv.2010.12.007.

Moo-Hyun, K., 2013. Spar Platforms: Technology and Analysis Methods. American Society of Civil Engineers, ISBN: 978-0784412091, ISBN-13.

Mutlu, S.B., Jorgen, F., 2003. Hydrodynamics Around Cylindrical Structure, revised ed. World Scientific Publishing Co. Pte. Ltd.

Nagavinothini, R., Chandrasekaran, S., 2019. Dynamic analyses of offshore triceratops in ultra-deep waters under the wind, wave, and current. Structures 20, 279–289. Available from: https://doi.org/10.1016/j.istruc.2019.04.009.

Ngo, Q.H., Hong, K.S., 2010. Sliding-mode anti-sway control of an offshore container crane. IEEE/ASME Trans. Mechatron. 17 (2), 201–209. Available from: https://doi.org/10.1109/TMECH.2010.2093907.

Ngo, Q.H., Hong, K.S., 2012. Adaptive sliding mode control of container cranes. IET Control Theory Appl. 6 (5), 662–668. Available from: https://doi.org/10.1049/iet-cta.2010.0764.

Nguyen, D.H., Nguyen, D.T., Quek, S.T., Sørensen, A.J., 2010. Control of marine riser end angles by position mooring. Control Eng. Pract. 18 (9), 1013–1021. Available from: https://doi.org/10.1016/j.conengprac.2010.05.001.

Niedzwecki, J.M., Liagre, P.Y., 2003. System identification of distributed-parameter marine riser models. Ocean Eng. 30 (11), 1387–1415. Available from: https://doi.org/10.1016/S0029-8018(02)00110-5.

Odijie, A.C., Wang, F., Ye, J., 2017. A review of floating semisubmersible hull systems: column stabilized unit. Ocean Eng. 144, 191–202. Available from: https://doi.org/10.1016/j.oceaneng.2017.08.020.

Owen, J.C., Masa, B., 2003. Experimental studies of passive control of vortex-induced vibration. J. Fluids Struct. 15, 597–605.

Pankanin, G.L., Berliński, J., Chmielewski, R., 2000. Karman vortex street visualization. In: Proceedings of International Symposium on Flow Visualization, vol. 239. Edinburgh, pp. 26–29.

Park, H., Bernitsas, M.M., Kumar, R.A., 2012. Selective roughness in the boundary layer to suppress flow-induced motions of circular cylinder at 30,000 $\leq$ Re $\leq$ 120,000. J. Offshore Mech. Arct. Eng 134 (4), 041801. Available from: https://doi.org/10.1115/1.4006235.

Park, H.R., Kumar, R.A., Bernitsas, M.M., 2013. Enhancement of flow induced motions of rigid circular cylinder on springs by localized surface roughness at $3 \times 104 \leq \mathrm{Re} \leq 1.2 \times 105$. J. Ocean Eng. 72 (1), 403–415. Available from: https://doi.org/10.1016/j.oceaneng.2013.06.026.

Parkinson, G.V., 1989. Phenomena and modeling of flow-induced vibrations of bluff bodies. J. Prog. Aerosp. Sci. 26, 169–224. Available from: https://doi.org/10.1016/0376-0421(89)90008-0.

Phifer, E.H., Kopp, F., Swanson, R.C., Allen, D.W., Langner, C.G., 1994. Design and installation of auger steel catenary risers. In: Offshore Technology Conference. doi:10.4043/7620-MS.

Qiao, D., Wu, F., Ou, J., 2014. Coupled Analysis of a Semisubmersible platform with two types of mooring systems. In: Proceedings of the 24th International Ocean and Polar Engineering Conference.

Rajan, T.P.D., Pillai, R.M., Pai, B.C., 2010. Characterization of centrifugal cast functionally graded aluminum-silicon carbide metal matrix composites. Mater. Charact. 61 (10), 923–928. Available from: https://doi.org/10.1016/j.matchar.2010.06.002.

Resvanis, T.L., Vandiver, J.K., Jhingran, V., Liapis, S., 2012. Reynolds number effects on the vortex-induced vibration of flexible marine risers. In: Proceedings of the ASME 2012 31st International Conference on Ocean, Offshore and Arctic Engineering. doi:10.1115/OMAE2012-83565.

Ricardo, A.G., 2008. A Study on Passive Methods of Vortex Induced Vibrations (Master of Science thesis). Massachusetts Institute of Technology, Department of Aeronautics Astronautics.

Rodolfo, T.G., Rosetti, G.F., Fujarra, A.L.C., 2011. Experimental comparisons to assure the similarity between VIM (vortex-induced motion) and VIV (vortex-induced vibration) phenomena. In: Proceedings of the ASME 2011 30th International Conference on Ocean, Offshore and Arctic Engineering. doi:10.1115/OMAE2011-49011.

Rustad, A.M., Larsen, C.M., Sørensen, A.J., 2008. FEM modelling and automatic control for collision prevention of top tensioned risers. Mar. Struct. 21 (1), 80–112. Available from: https://doi.org/10.1016/j.marstruc.2007.04.003.

Salama, M.M., Johnson, D.B., Long, J.R., 1998. Composite production riser testing and qualification. SPE Prod. Facil. 13 (03), 170–177. Available from: https://doi.org/10.4043/8432-MS.

Salama, M.M., Stjern, G., Storhaug, T., Spencer, B., Echtermeyer, A., 2002. The first offshore field installation for a composite riser joint. In: Offshore Technology Conference. doi:10.4043/14018-MS.

Sarpkaya, T., 1977. In-line and transverse forces on cylinders in oscillatory flow at high Reynolds numbers. J. Ship Res. 21, 200–216.

Sarpkaya, T., 1978. Fluid forces on oscillating cylinders. ASCE J. Waterw. Port. Coast. Ocean Div. 104, 275–290.

Sarpkaya, T., 2004. A critical review of the intrinsic nature of vortex-induced vibrations. J. Fluids Struct. 19, 389–447.

Sauer, C.W., Sexton, J.B., Sokoll, R.E., Thornton, J.M., 1996. Heidrun TLP titanium drilling riser system. In: Offshore Technology Conference. doi:10.4043/8088-MS.

Shen, C., Pan, Z., Cuiuri, D., Dong, B., Li, H., 2016a. In-depth study of the mechanical properties for Fe_3Al based iron aluminide fabricated using the wire-arc additive manufacturing process. Mater. Sci. Eng.: A 669, 118–126. Available from: https://doi.org/10.1016/j.msea.2016.05.047.

Shen, C., Pan, Z., Cuiuri, D., Roberts, J., Li, H., 2016b. Fabrication of Fe-FeAl functionally graded material using the wire-arc additive manufacturing process. Metall. Mater. Trans. B 47 (1), 763–772. Available from: https://doi.org/10.1007/s11663-015-0509-5.

Shishkovsky, I., Missemer, F., Smurov, I., 2012. Direct metal deposition of functionally graded structures in the Ti-Al system. Phys. Procedia 39, 382–391. Available from: https://doi.org/10.1016/j.phpro.2012.10.052.

Silvestre, N., 2008. Buckling behavior of elliptical cylindrical shells and tubes under compression. Int. J. Solids Struct. 45 (16), 4427–4447. Available from: https://doi.org/10.1016/j.ijsolstr.2008.03.019.

Sparks, C.P., 2007. Fundamentals of Marine Riser Mechanics: Basic Principles and Simplified Analyses. PennWell Books, ISBN: 978-1-59370-070-6, ISBN.

Srinivasan, N., Chakrabarti, S., Radha, R., 2006. Response analysis of a truss-pontoon semisubmersible with heave-plates. J. Offshore Mech. Arct. Eng. 128 (2), 100–107. Available from: https://doi.org/10.1115/1.2185679.

Stansberg, C.T., 2008. Current effects on a moored floating platform in a sea state. In: ASME 2008 27th International Conference on Offshore Mechanics and Arctic Engineering. American Society of Mechanical Engineers Digital Collection, pp. 433–444. https://doi.org/10.1115/OMAE2008-57621.

Sunil, D.K., Mukhopadhyay, M., 1995. Free vibration of semisubmersibles: a parametric study. Ocean Eng. 22 (5), 489–502. Available from: https://doi.org/10.1016/0029-8018(94)00012-V.

Susan, B.S., Vandiver, J.K., Larsen, C.M. Lie, H., 2008. Reynolds number dependence of flexible cylinder VIV response data. In: Proceedings of the ASME 27th International Conference on Offshore Mechanics and Arctic Engineering. OMAE2008-57045.

Tennyson, R.C., Booton, M., Caswell, R.D., 1971. Buckling of imperfect elliptical cylindrical shells under axial compression. AIAA J. 9 (2), 250–255. Available from: https://doi.org/10.2514/3.6159.

Tim Finnigan, T.D., Irani, M.B., Van Dijk, R.R.T., 2005. Truss spar VIM in waves and currents. In: Proceedings of 24th International Conference on Offshore Mechanics and Arctic Engineering. doi:10.1115/OMAE2005-67054.

Trim, A.D., Braaten, H., Lie, H., Tognarelli, M.A., 2005. Experimental investigation of vortex-induced vibration of long marine risers. J. Fluids Struct. 21 (3), 335–361. Available from: https://doi.org/10.1016/j.jfluidstructs.2005.07.014.

Übeyli, M., Balci, E., Sarikan, B., Öztaş, M.K., Camuşcu, N., Yildirim, R.O., et al., 2014. The ballistic performance of SiC–AA7075 functionally graded composite produced by powder metallurgy. Mater. Des. (1980–2015) 56, 31–36. Available from: https://doi.org/10.1016/j.matdes.2013.10.092.

Umar, A., Ahmad, S., Datta, T.K., 2004. Stability analysis of a moored vessel. J. Offshore Mech. Arct. Eng. 126 (2), 164–174. Available from: https://doi.org/10.1115/1.1710873.

Wang, K., Er, G.K., Iu, V.P., 2018. Nonlinear vibrations of offshore floating structures moored by cables. Ocean Eng. 156, 479–488. Available from: https://doi.org/10.1016/j.oceaneng.2018.03.023.

Webster, W.C., 1995. Mooring induced damping. Ocean Eng. 22 (6), 571–591. Available from: https://doi.org/10.1016/0029-8018(94)00027-5.

Williamson, C.H.K., 1988. The existence of two stages in the transition to three dimensionality of a cylinder wake. J. Phys. Fluids 31 (11), 3165–3168.

Williamson, C.H.K., Govardhan, R., 2004. Vortex induced vibrations. J. Annu. Rev. Fluid Mech. 36, 413–455.

Williams, S.W., Martina, F., Addison, A.C., Ding, J., Pardal, G., Colegrove, P., 2016. Wire + arc additive manufacturing. Mater. Sci. Technol. 32 (7), 641–647. Available from: https://doi.org/10.1179/1743284715Y.0000000073.

Wu, Y., Wang, T., Eide, Ø., Haverty, K., 2015. Governing factors and locations of fatigue damage on mooring lines of floating structures. Ocean Eng. 96, 109–124. Available from: https://doi.org/10.1016/j.oceaneng.2014.12.036.

Xu, J., He, M., Bose, N., 2009. Vortex modes and vortex-induced vibration of a long, flexible riser. Ocean Eng. 36 (6–7), 456–467. Available from: https://doi.org/10.1016/j.oceaneng.2009.01.010.

Xu, S., Ji, C.Y., Soares, C.G., 2018. Experimental and numerical investigation a semi-submersible moored by hybrid mooring systems. Ocean Eng. 163, 641–678. Available from: https://doi.org/10.1016/j.oceaneng.2018.05.006.

Xue, H.X., Tang, W.Y., Zhang, S.K., 2008. Simplified prediction model for vortex-induced vibrations of top tensioned risers. China Ocean Eng. 22 (2), 291–302.

Xue, H., Tang, W., Qu, X., 2014. Prediction and analysis of fatigue damage due to cross-flow and in-line VIV for marine risers in non-uniform current. Ocean Eng. 83, 52–62. Available from: https://doi.org/10.1016/j.oceaneng.2014.03.023.

Xue, X., Chen, N.Z., Wu, Y., Xiong, Y., Guo, Y., 2018. Mooring system fatigue analysis for a Semi-submersible. Ocean Eng. 156 (550–563). Available from: https://doi.org/10.1016/j.oceaneng.2018.03.022.

Yan, J., Qiao, D., Ou, J., 2018. Optimal design and hydrodynamic response analysis of deep-water mooring system with submerged buoys. Ships Offshore Struct. 13 (5), 476–487. Available from: https://doi.org/10.1080/17445302.2018.1426282.

Yang, Y., Chen, J.X., Huang, S., 2016. Mooring line damping due to low frequency superimposed with wave frequency random line top-end motion. Ocean Eng. 112, 243–252. Available from: https://doi.org/10.1016/j.oceaneng.2015.12.026.

Yeo, J.G., Jung, Y.G., Choi, S.C., 1998. Design and microstructure of ZrO2/SUS316 functionally graded materials by tape casting. Mater. Lett. 37 (6), 304–311. Available from: https://doi.org/10.1016/S0167-577X(98)00111-6.

Yuan, Z., Ji, C., Chen, M., Zhang, Y., 2011. Coupled analysis of floating structures with a new mooring system. In: Proceedings of the 30th International Conference Ocean, Offshore and Arctic Engineering. pp. 489–496. https://doi.org/10.1115/OMAE2011-49597.

Yuan, H., Li, J., Shen, Q., Zhang, L., 2012. In situ synthesis and sintering of ZrB2 porous ceramics by the spark plasma sintering–reactive synthesis (SPS–RS) method. Int. J. Refract. Met. Hard Mater. 34, 3–7. Available from: https://doi.org/10.1016/j.ijrmhm.2012.01.007.

Zdravkovich, M.M., 1981. Classification of various aerodynamic and hydrodynamic means for suppressing vortex shedding. J. Wind Eng. Ind. Aerodyn. 7, 145–189.

Zhang, J., Tang, Y., 2015. Fatigue analysis of deep-water risers under vortex-induced vibration considering parametric excitations. J. Coast. Res. 73 (sp1), 652–660. Available from: https://doi.org/10.2112/SI73-113.1.

Zhang, H.T., Feng, J.C., He, P., Zhang, B.B., Chen, J.M., Wang, L., 2009. The arc characteristics and metal transfer behavior of cold metal transfer and its use in joining aluminum to zinc-coated steel. Mater. Sci. Eng.: A 499 (1–2), 111–113. Available from: https://doi.org/10.1016/j.msea.2007.11.124.

Zhu, J., Collette, M., 2017. A Bayesian approach for shipboard lifetime wave load spectrum updating. Struct. Infrastruct. Eng. 13 (2), 298–312. Available from: https://doi.org/10.1080/15732479.2016.1165709.

Index

Note: Page numbers followed by "*f*" and "*t*" refer to figures and tables, respectively.

CPI Antony Rowe
Eastbourne, UK
October 30, 2020